The Mountain

AN AUTHORITATIVE GUIDE TO THE TABLE MOUNTAIN CHAIN

Edited by Dr Douglas Hey

TAFELBERG

Cover photo © Mark van Aardt

Frontispiece:
Table Mountain covered by its spectacular tablecloth.
(*Photo: Zelda Wahl*)

Tafelberg Publishers Limited, 28 Wale Street, Cape Town 8001

Typography and cover design by G&G Design
Set in 10 on 12 pt Plantin
Printed and bound by National Book Printers, Goodwood, Cape
First edition 1994

ISBN 0 624 03323 6

Acknowledgements

We wish to express our sincere appreciation to all those who assisted in the compilation of this book, which, it is hoped, will promote the conservation of Table Mountain.

An important contribution was made by the many photographers who generously donated material on the fauna, flora, geological, historical and scenic features of the mountain, including the few anonymous photographers who regrettably could not be mentioned by name. To all of them we say thank you.

THE PUBLISHER

Preface

As a diversified transport group and a major employer, Transnet is fully committed to help in building a better future for South Africa. In order to meet the challenges of the future, Transnet lays stress on enhancing the quality of its services and on the development and training of its human resources. It acknowledges, however, that it not only has a business and human responsibility, but that as a key player in the South African economy, it also has an obligation to preserve the environment in which it operates. Based on this holistic approach, Transnet's environmental policy strives for harmony and synergy between its operational activities and the preservation of our cultural, historical and natural environment.

Today's challenge particularly lies in creating an acceptable standard of living for all South Africans whilst at the same time conserving or restoring the viability of the natural environment – and thus the very basis of human life. Transnet is firmly committed to meet this challenge.

Not only has Transnet played a pioneering role in Cape Town from the earliest times; it has also, through its subsidiary Victoria & Alfred Waterfront, masterminded the restoration of the historic link between the City of Cape Town and the sea. The maintenance of full commercial harbour activities together with the rehabilitation of individual buildings and facilities adaptable for diverse uses are further significant examples of integrated conservation.

The incomparably beautiful mountain range, with the Kirstenbosch National Botanic Gardens on its slopes, provides a sanctuary for a great variety of the Cape's unique fauna and flora, thereby contributing to the conservation of what Linnaeus described as "a paradise on earth . . . more valuable than the wealth of Solomon or Croesus".

Since a vast area of the natural fynbos has already been destroyed as a result of development and negligent human interference of one kind or another, the conservation of this natural heritage deserves the unanimous support of every South African as part of our commitment to the future and for the sake of generations to come.

Transnet is proud to be associated with this authoritative and definitive work on Table Mountain and the conservation of this famous landmark.

Contents

Foreword

This book is a unique and authoritative collaborative effort to highlight the plight of Table Mountain, one of the world's best-known and best-loved landmarks.

In recent years there has been a marked awakening in awareness of Table Mountain's importance and the need for its proper utilisation. In a little more than 300 years the mountains of the Cape Peninsula have undergone a remarkable transformation. Early explorers recorded that lion, elephant, rhino, buffalo and eland inhabited its slopes. The last lion was killed in 1802. By the end of the last century large parts of the mountain were barren because trees were chopped down for firewood and timber. Today more than 300 footpaths lead to the mountain's summit and more than 10 million people have visited the top using the cableway.

Although Table Mountain is under increasing pressure, it remains a remarkable natural and recreational asset. Nowhere else in the world is an area of such scenic beauty and biological diversity enclosed within the perimeter of a fast-growing city. This book provides essential information on a subject which should be more fully understood by all, namely the urgent need for the conservation of Table Mountain and the mountains of the Cape Peninsula.

The Mountain, with its authoritative chapters on Table Mountain's human history, geology, fauna and flora, humán usage and conservation, will help foster the ethic of stewardship which this unique natural and recreational area so desperately needs.

JOHN HANKS
Chief Executive, Southern African Nature Foundation

Introduction

From the earliest times, Table Mountain has been recognised by navigators as a remarkable geographical feature, and Table Bay as one of the most scenic anchorages in the world. Scientists have described the mountain as a unique natural wonder due to the rich and diverse flora growing in a setting of majestic proportions and extraordinary beauty. It is a unique natural asset, regarded with pride by many citizens of our country who seek recreation, physical and spiritual regeneration and inspiration on its slopes. In an address on Table Mountain General J C Smuts said:

"We, as a nation, valuing our unique heritage, should not allow it to be spoiled and despoiled, and should look upon it as among its most sacred possessions, part not only of the soil, but of the soul of South Africa. For centuries to come, while civilisation lasts on this subcontinent, this national monument should be maintained in all its natural beauty and unique setting. It should be symbolic of our civilisation itself, and it should be our proud tradition to defend it to the limit against all forces of man or nature to disfigure it."

The first humans to utilise the mountains were probably the Khoikhoi tribes living on the Cape Peninsula, as no traces of San (Bushman) art or artefacts have been found so far south. Apart from hunting and trapping and gathering fruits and bulbs, the Khoi herdsmen would have grazed their livestock on the lower slopes. The Khoikhoi developed the practice of burning the veld periodically to provide green grazing for their animals. Early seafarers rounding the Cape commented on the dense clouds of smoke enveloping Table Mountain.

Since the first Dutch settlement at the Cape in 1652, the Table Mountain complex has been exploited, abused, burnt, eroded, vandalised with graffiti and despoiled. Within a few decades of the establishment of the settlement, the mountain was stripped of readily available timber for shipbuilding, housing and furniture-making, and the woodcutters moved on to Hout Bay. Despite legislation to control the exploitation of the forests, this area was also cleared of usable trees by 1772. At the time, wood was the sole source of fuel for the settlement and passing ships, and vast quantities were collected. The mountain slopes were combed daily for firewood and fuel for charcoal-making and the limekilns. The bakers were also major consumers of wood.

As the settlement grew and the demand for firewood increased, the wood collectors moved higher and higher up the mountain, and when this source was exhausted, out onto the Cape Flats and further afield. Carl Thunberg, the Swedish botanist and traveller, wrote in 1772: "There are no forests in the vicinity of the town, except a few remnants high up in the clefts of the mountains; wood, which is used in the kitchen, is both dear and scarce." As our trees and shrubs are slow growers, the wholesale collection of timber and firewood had a marked effect on the appearance of the mountain slopes.

The growing public reaction to the disfiguring of Table Mountain stimulated the authorities to launch an afforestation programme to "clothe these barren slopes", to green the mountain and also provide much-needed timber. A variety of fast-growing trees were introduced from Europe and Australia. This culminated in the establishment of the Devil's Peak forestry station. The work of Mr F Jarman, the forester responsible for this project, is commemorated on a bronze plaque near the King's Blockhouse on the slopes below Devil's Peak. Unfortunately a few of the species of trees selected, such as the pinaster pine and Australian acacias, have proved invasive and most undesirable introductions and a major threat to the indigenous vegetation.

This book is an endeavour to present a popular account of the history, geology, fauna and flora, and human usage of the mountains of the Cape Peninsula, emphasising the importance of the conservation of this unique natural resource, not only as habitat for the fauna and flora but also for outdoor recreation and tourism.

Conservation means sound and wise management of nature including utilisation, but only within the natural potential of the resource. The major threat to nature areas today is the increasing pressure of human numbers, leading to development and excessive usage, and this holds equally true for the Peninsula mountains. The conservation of this priceless national asset is the subject of the concluding chapter of the book.

DR DOUGLAS HEY
Rondebosch

Human history of Table Mountain

DR EDNA BRADLOW

"Our misery and sorrows now being at their height, and there being no way out as far as human eyes could see, God Almighty whose goodness towards us poor men endures . . . now to the great rejoicing of us all gave the sight of the high land of the Cape de Bona Esperance lying at 34 degrees south latitude and 38 degrees of longitude. It is impossible to describe in words with what joy and happiness this was seen by our crew: the lame, crippled and even those who could hardly rise from their berths came on deck to see the land."

These words, written in 1640 by a Dutch ship's surgeon, Nicolaus de Graaff, symbolise a basic aspect of the protean relationship between human beings and Table Mountain: the great relief felt by European voyagers when, after months at sea, they saw the first recognisable landmark – what a Dutch navigator, Joris van Spilbergen, described in 1601 as "the high hill, flat above and square like a table, and visible for nine or ten miles to seaward" – with the welcoming, eponymous bay at its feet.[1]

Relief was accompanied by awe. Two centuries later, William Burchell, the great English traveller and natu-

Some of the early artists' renderings of Table Mountain were extremely fanciful.

ralist, was to articulate more imaginatively the emotions felt on approaching Table Bay by sea. "As we advanced nearer along the shore, the mountains displayed an imposing grandeur which mocked the littleness of human works: buildings were too small to add a feature to the scene; too insignificant either to adorn or to disturb the magnificence of nature."[2]

"If it be clere," wrote Alexander Childe, an Englishman, in 1616, ". . . you shall see it ryse like a tableland and other ragged heiles from it to the southwards and to the northwards."[3] Estimates of the mountain's height varied wildly, from Thomas Herbert's 11 860 ft in 1627 to the more accurate 3 300 ft scientifically arrived at by a fellow-traveller of the English factor Peter Mundy in 1634.[4]

The "rugged hills" running from east to west and known today as Devil's Peak, Lion's Head and Signal Hill were given other descriptive names by the first Europeans to see them. Thus Devil's Peak was known variously by the English as "the other hill", "the altar", "King Charles' Mount" (which was the name originally given to Signal Hill), and by the Dutch as "Windberg". Peter Mundy described it as "being certaine rocks on the Topp of a hill, resemblinge a castle afar off".[5] In 1795 at the start of the British occupation, three blockhouses were built on Devil's Peak to protect the settlement from attack.

While the English persisted in calling Lion's Head the Sugarloaf, by 1627 the Dutch had recognised its resemblance to a "lying lion" and named it Lion's Hill. When in 1620 Fitzherbert and Shillinge claimed Table Bay on behalf of the English sovereign, they named what is today Signal Hill, "King James his Mount". Subsequently, it was seen as part of the "lying lion" and called Lion's Rump.

Its usage as an outlook for monitoring ships' movements dates from early in the Dutch period. As Francois Valentyn, the Dutch pastor who visited Cape Town in the late 17th century, noted, "on top of it a flagstaff stands, near which one or two sentries stay . . . who at once hoist the flag when they see any ships far out at sea, which they can do when fully 10 or 12 miles away, and fire a gun set there, as many times as there are ships, to give news of them to those in the Fort."[6] Signal Hill's

Bowler's "Table Bay and Mountain from Blouberg" with the familiar "rugged hills" described by the early travellers.

modern name possibly dates from soon after 1878 when a permanent signal station was built on the hill.[7]

Table Bay lay on what was to become, from the 16th century onwards, one of the busiest shipping lanes in the world. Consequently, Europeans became familiar, through the impressions of artists, with the distinctive outline of Table "Hill". Many of these draughtsmen had never seen the mountain and worked up their pictures from sailors' accounts, so that some of the early renderings were fanciful in the extreme. But even Wouter Schouten, an early artist who did visit the Cape as a ship's surgeon in the 1660s, grossly distorted the front face of the mountain.

Perhaps the most accurate of the early pictures of Table Mountain was that done in 1772 by Captain Cook's artist, William Hodges, whose responsibility was primarily to record visually the weather conditions and topographical features encountered by the expedition. Hodges was particularly interested in the effects of light, and this is apparent in his beautiful picture "A View of the Cape of Good Hope", housed today in the National Maritime Museum, Greenwich.

As the roughly halfway mark along the important sea lane linking Europe to the Indian Ocean and the Far Eastern trading network, Table Bay's advantages for victualling passing ships were obvious. Fresh water, as we shall see, was readily available. The procurement of fresh meat depended pre-eminently, however, upon the indigenous peoples whom the Europeans encountered, probably for the first time in 1503 when the Portuguese captain Antonio de Saldanha stepped ashore in Table Bay. By the time of Van Riebeeck's arrival, the Peninsula Khoikhoi numbered between 4 000 and 8 000 people, organised into small groups. Livestock were their most valuable asset and they consequently competed for the pasturage around Table Mountain. The historian Richard Elphick locates their favourite camping sites as the mouth of the Salt River; behind the mountain in the Hout Bay Valley; the area near the modern Chapman's Peak road; and present-day Bishopscourt.[8] Thus Nicholas Downton, an English mariner, wrote in 1610 that his commander "desired to walk about the Table to see if we could see anye cattle, that we might by any means buy for our reliefe".

From the moment the Dutch stepped on shore in Table Bay to establish a refreshment station, the mountain dominated their lives, starting with the very locations in which it would virtually compel them to settle. Their intention of using the Cape merely as an outpost of an extended commercial empire initially determined the restricted dimensions of "het Kaapsche vlek" (literally the "Cape stain", or Cape settlement).

Watercolour of shipping in Table Bay, 1752, by Mattheus Sager. (*Photo: Africana Museum*)

Plates with the image of Table Mountain were decorated by Chinese artists for the Dutch East India Company.

W M Craig's "Cape of Good Hope", 1806, from which the aquatint by E. Orme was engraved.

The station began with a fort and a garden. The first fort for the defence of the small settlement was built of rammed earth on the site of the present Parade. It was replaced in 1665 with a pentagonal fortification of dressed stone with bastions at each corner, on a site south-east of the old fort. It still stands today, the oldest and historically the most interesting building in the country.[9]

The settlement and the ships obtained their water from a strong mountain stream, the Fresh River (Capel Sluit), which flowed past the fort to debouch into Table Bay. Van Riebeeck's successor, Zacharias Wagenaar, ordered the construction of a wall, channel and sluices to facilitate the filling of water casks.

The first garden was laid out between the fort and the Fresh River and was gradually extended across the river and up towards the mountain. Thus the initial expansion beyond the confines of the protective fort followed the rich alluvial deposits of the watercourses running down the mountain. Rust en Vreugd, Oranjezicht, Rheezicht and Leeuwenhof were the best known estates subsequently established on these intensively cultivated anterior slopes of Table Mountain. As late as 1853, when the young Swede J F Victorin visited the Cape, the scene immediately below the north face could still be described as "pastoral".[10]

During the 18th century, however, the town had been growing *pari passu*, "at the foot and along the first slope of the Lion Hill . . . Hottentot (i.e. Riebeeck) Square being considerably up the ascent".[11] This was the artisan quarter, whose dwelling houses were occupied by the Cape Muslims after the termination of slavery. They buried their dead where they had lived; the first Muslim burial ground was also on the lower slopes of Signal Hill.

These were future developments. When Van Riebeeck sought to grow grain in Table Valley, he found that the prevailing winds destroyed his crops. Consequently he turned to the sheltered land below the eastern slopes of the mountain which were already being

denuded of their timber within the first decade of the Dutch settlement. The allocation in 1657 of free burgher farms along the Liesbeek River transformed the former Khoikhoi grazing ground into arable land producing cereal crops. Many of the Khoikhoi went into the service of the Dutch, but a few individuals withdrew to, and remained in, the remoter areas of the Table Mountain massif.

Van Riebeeck himself acquired an estate at Bosheuvel (the latter-day Bishopscourt) where he planted the first vineyard in South Africa. The Vereenigde Oost-Indische Compagnie (VOC) established "magnificent orchards"[12] at 't Ronde Doornbosjen (Rondebosch), "also called Rustenburg".[13] Kirstenbosch, adjoining Bosheuvel, was another of the company's agricultural enterprises.[14]

In an attempt to protect the growing settlement against stock thieves, Van Riebeeck planned a "defensive fence" of fast-growing trees and shrubs, so dense that cattle or sheep could not be driven through it. A wide hedge of bitter almonds (*Brabejum stellatifolium*), brambles and thorn bushes was duly planted on the boundary of the settlement. The remnants of this can be seen in Kirstenbosch and on Wynberg Hill and have been declared a national monument.

In 1772 Anders Sparrman, the Swedish botanist, standing on the eastern summit of the mountain, saw below him "several neat, compact farms scattered up and down".[15] But the intensive mixed agriculture envisaged by Van Riebeeck for those eastern foothills had not materialised.[16] Instead the free burghers became pastoralists, and with their need for more grazing the settlement expanded westwards into the Hout Bay Valley and southwards along the foot of the Steenberg and Constantia mountains. In the 1680s the cattle farmers began to be ousted from these areas by the estates of the great agriculturists. Governor Simon van der Stel entertained "nobly" on his Constantia estate[17] which subsequently became famous for its export wine, "extremely sweet, agreeable and luscious . . . and only fit for the dessert, as on account of its sweetness, if drunk in abundance, it lies heavy on the stomach"[18]; Alphen, granted in 1714 to Theunis van Schalkwyk, possessed a substantial house and extensive grounds when Sparrman saw it in the 1770s.[19]

While Wynberg preserved its "elysian scenery"[20] well beyond the middle of the nineteenth century, the utilisation of space nearer the metropolis for residential purposes brought suburbia to the south-east side of the mountain. Already in William Burchell's time – 1810 – Rondebosch was "an assemblage of villas and gardens",[21] and this extension of the southern suburbs was accelerated by the construction of the suburban railway line.[22]

By the end of the century, the last of the great estates in this area was Groote Schuur. Cecil Rhodes bought it in 1891, adding to it "adjoining farms on the lower slopes and in the little valleys of Table Mountain, so that in the end he had an estate of some 1 500 acres".[23] The University of Cape Town was built on this estate, the first sod being turned by Otto Beit in 1920. The university chose two locations for its buildings. One, the site of the medical school, lay at the furthest edge of Rhodes's property; the other, the most level piece of ground available, is where the main campus is situated.

In 1741 the VOC had established a winter anchorage at Simon's (formerly Yselstein) Bay, protected by the Table Mountain massif from the destructive north-west gales; and gradually a small settlement developed there, hard up against the Simonsberg. A fort, built in 1743 at the company's cattle post and commanded originally by Wynand Muys, protected the entrance to the narrow sandy road that proceeded towards Simon's Bay between the mountain and the shore. By the end of the 19th century, "progress" had caught up with the fishing villages – Muizenberg and Kalk Bay – along this road, whose unique scenic combination of mountain and sea were making them desirable holiday places.

If what Trollope called "the cool side of the Table Mountain" was in the 19th century to become a desirable residential area for this very reason, the marine suburbs on the drier western side of the mountain were beginning to enjoy an equally enviable reputation. The Dutch and early English period had produced no great estates in this area (except perhaps Henry Alexander's

The kramat of Goolam Muhamed Soofi on Signal Hill, one of a number of such sacred burial places of Muslim leaders in the Cape Peninsula. (*Photo: Jose Burman*)

"Sea Point" extending up the lower slopes of Lion's Head). In the early years of the 19th century, however, there were about a dozen extensive residential properties in modern Green and Sea Point, which from the 1850s onwards were being divided into smaller building lots. Thus, as Anthony Trollope noted in 1877, the second road out of Cape Town led to "a second pleasant suburb and a second clustering together of villa residences". There was, however, also a third road, "too steep for wheels",[24] constructed by the military during Sir Harry Smith's governorship. It ran up modern Kloof Nek Road, over Kloof Nek and then descended, "skirting at a considerable height the rocky shores of Camps Bay until you finally reach the villa-covered heights of Sea Point and Green Point".[25] It was from this road that, during the American Civil War, Capetonians watched the Confederate raider Alabama capture a Federal prize off Camps Bay.

Thus by 1903 the whole area from Sea Point to Diep River – on the plains and hard up against the mountain – was residential; while Kalk Bay and Simon's Town, rising from their sandy localities and narrowly situated between sea and mountain, were also providing living space for Capetonians. The mountain, then, determined the settlement pattern of the early town; and having settled, its inhabitants used the mountain's resources in many different ways. Fresh water was responsible for the very existence of the town; for Saldanha Bay, a safer anchorage, lacked that precious commodity.

Many of the pre-Van Riebeeck accounts refer to the quality and quantity of the water in the two main streams – the Platteklip and the Capel Sluit (the Fresh River) – which were fed by smaller rivulets produced through the precipitation of moisture on the mountain's summit.[26] The English naval chaplain Edward Terry referred in 1616 to "a most delectable brook of pure good water, running hard by out of a mighty hill (called from its form the Table)". Thomas Herbert was repeating the almost prescriptive sentiments expressed by mariners when he wrote of tasting this "sweet and wholesome water . . . which after so long being at sea was a great refreshing to our scorcht Intrails".[27]

These two streams constituted the main water supply of central Cape Town almost until the end of the 19th century. In the 18th century, the cultivators on the lower mountain slopes used them for irrigating their crops during the long summer. They provided the motive power for several mills, one of which, just below the start of Platteklip Gorge, Burchell described as "the most elevated habitation in Table Valley".[28] More important was the provision to ships of fresh water, which was brought in wooden pipes from the mountain to the jetty near the Castle. These streams filled the decorative canals of the town until the early 19th century when the destruction of covering vegetation on the mountain reduced the flow of run-off water. Eventually, the canals deteriorated into stinking sewers choked with garbage, and were covered over.

With increasing urbanisation during the 18th century, public fountains were provided into which drinking-water from the foot of the mountain was piped. These fountains continued to exist well into the following century, as several of Thomas Bowler's pictures indicate.[29] Furthermore the "brooks that come leaping down from the Table Mountain" were a valuable resource for dozens of washerwomen, "lathering and wringing, banging and pounding our unfortunate garments".[30]

Table Mountain's water remained one of the Peninsula's most important assets. By the start of the 20th century, two of the three systems providing the city and some of its suburbs with water depended on rainfall collected in the catchment areas on the summit and in Orange Kloof, and stored in reservoirs (such as Molteno, Woodhead and Newlands), partly built by Cornish stonemasons. Even the two isolated southern Peninsula municipalities – Kalk Bay and Simon's Town – depended for their water supply on the run-off water from the Table Mountain range.[31]

This bounty was not, however, wholly beneficial. The flow of water over the steep rocks and grassy slopes of the mountain created recurrent drainage problems during severe winter storms, necessitating the construction of an extensive underground system for carrying off the unstored rainfall.

Wood was as essential as water for ensuring the survival of Van Riebeeck's settlement. The building of a fort, jetty and houses required timber; cooking and heating (including the firing of bricks) depended upon the availability of firewood in the absence of coal deposits. In 1620, Augustin de Beaulieu, a French admiral, had found in Table Bay Valley "a great many trees, including some from which planks a foot wide and 18 to 20 feet long could be cut". These were large silvertrees; but de Beaulieu must have overestimated their number. For within the first year of the Dutch settlement Van Riebeeck was forced to explore, and exploit, the kloofs on the eastern and southern sides of the mountain, where the high rainfall and cool summers encouraged the growth of larger trees.[32]

To convey the timber to the settlement, Van Riebeeck constructed the "father of South African highways", the "wagen pad na 't bos".[33] It rounded Devil's Peak, proceeded past Rondebosch, and wound along the route of present-day Newlands Avenue to the kloof above the company's abundant forests at Post 't Paradys and Kirstenbosch (the latter, Van Riebeeck noted in his journal in August 1653, "as full of timber as the large and small forests outside of Haarlem").

Initially, the Europeans were unaware of the slow growth of our indigenous trees and felled them indiscriminately. The exhaustion of these forests, particularly on the lower slopes of the mountain, led the Dutch to extend the woodcutters' road over Constantia Nek into the Hout Bay Valley, where, as its name indicates, even denser forests were found and exploited.

Van Riebeeck soon became concerned at the manner in which these forests were being "recklessly destroyed". In 1657 Leendert Cornelisz was appointed sole woodcutter for the area where Kirstenbosch now stands, and where the name Leendertsbos marks the site of his store. The following year, in October 1658, a decree was promulgated which further limited free access to the forests and controlled their exploitation; but owing to lack of enforcement, these measures were ineffective.

In the 1770s the Swedish botanist and traveller Carl Thunberg commented on the absence of large forests anywhere near the town. However, some of the indigenous trees that grew along the back of the mountain, such as boekenhout, ironwood and assegai, have been and are still, found occasionally in the less accessible kloofs. In April 1936 General J C Smuts came across a few of these "survivors of the primitive forests".[34]

The first efforts to augment the dwindling timber resources of the Cape with faster-growing trees date back to 1655 when young oaks and firs were shipped from Holland. This tree-planting programme was continued by subsequent governors; Simon van der Stel, for example, planted hundreds of oaks at Hout Bay and on the Steenbergen.

The first systematic, large-scale plantings of exotic

Cape Town from "near the Amsterdam Battery" by T W Bowler, showing the bare north face of the mountain, denuded of its large trees.

trees on the Peninsula mountains, however, date from 1875 with the appointment of J Storr Lister as superintendent of plantations. He was responsible for the establishment of the historic nurseries and arboretum at Tokai (later to become the site of the first school of forestry), and for the Oude Kraal nurseries on the lower plateau of Table Mountain. A remarkable variety of trees were propagated with a view to afforesting the Peninsula mountains. As planting with seedlings was considered too slow, foresters resorted to broadcasting the seed of pines and oaks, sowing up to 1,5 tons per annum over many years. Below the King's Blockhouse on Devil's Peak there is a plaque in memory of the forester Frank Jarman which reads: "He found these barren stony slopes treeless and left them covered with forest."

Originally the proletarianized Khoikhoi were hired as casual firewood-gatherers. But with the introduction in 1658 of slaves, and because wood was so scarce, and hence expensive, the latter were permanently employed on this essential but arduous task. Slaves regularly scoured the mountain for firewood, either for their masters' domestic use or as a saleable commodity. Thunberg listed the varieties of small indigenous trees and bushes used for cooking; "the stems and roots of the Protea grandiflora, conocarpus, speciosa, hirta, mellifera and argentea, a few species of Erica, and some sorts of Brunia".[35]

Woodcutting under licence in the valley between Devil's Peak and Lion's Head was introduced in the 18th century, and this form of control was subsequently extended to the top of the mountain and along the massif. Nevertheless both Sparrman and Thunberg noted that the consumption of wood was fast exceeding the supply. It was obviously difficult to limit the owners' demands and therefore to control the activities of the slaves, who preferred the more accessible areas from which they brought huge loads on poles slung across their shoulders.[36]

They knew the mountain intimately, using paths untrodden by any other feet.[37] Throughout Cape history there are instances of runaway slaves who used the remoter kloofs as their refuge, while simultaneously maintaining communication with their fellows on the mountain or with the inhabitants of the sleazier urban quarters below. They lived in groups of up to fifteen members[38] in "the holes and clefts of the hill", where, according to Robert Semple, they "often set fire accidentally, or with design, to the brush and underwood which grows along the face of the hill".[39]

Table Mountain continued to support this exiguous population of socially marginal people, who sheltered in the caves situated throughout the range. Joshua Penny, an American sailor who escaped from British impressment after the Battle of Muizenberg, spent 14 months in such a cave without seeing another human being. "I never enjoyed life better than while I lived among the ferocious animals of Table Mountain," he later wrote, "because I had secured myself against the more savage English."[40] Today, the spiritual descendants of Penny and the refugee slaves are the "bergies", who with the familiar sacks on their backs scavenge among the rubbish of the affluent homes below the mountain face.[41]

In tracing the human history of Table Mountain, one must consider the use made by humans of the mountain's other material phenomena and collective objects. On Table Mountain, as elsewhere, man's innate greed has surfaced occasionally. As early as 1615 an English visitor was sending home samples of Table Mountain rock for assay[42]; and throughout the 19th century sporadic but mercifully ineffective attempts were made to exploit the small mineral deposits found on the Peninsula range. Manganese was mined for some time above Hout Bay; gold-bearing pyrites were discovered on Lion's Head but proved inadequate for profit; tin was found on the slopes of Devil's Peak but similarly the amount lacked commercial potential; while silver was unsuccessfully mined on the Steenberg Plateau above Tokai, just off the present-day Ou Kaapse Weg.

But the articulation of people and mountain has not been characterised by greed; rather, the mountain has primarily been a source of inspiration and pleasure, even of intellectual stimulation. In the pre-pollution days of Cape Town, the very height of Table Mountain seemed to make it a suitable vantage point for astronomical calculations; so, at least, the Cape colonial authorities believed. In 1821, when the first Astronomer Royal, the Reverend Fearon Fallows, was looking for a suitable hill on which to establish his observatory, they offered him a site on Table Mountain, "considering that proximity to the stars was a sine qua non".[43] Fallows climbed Table Mountain and measured its height with a barometer; for good measure, he sent home various natural objects collected on the mountain.

The most famous astronomical memorial associated with Table Mountain is, however, Maclear's Beacon, which commemorates the name of Sir Thomas Maclear, third incumbent at the observatory. His activities also link the mountain to the venerable French astronomer, the Abbé Nicolas de Lacaille, who arrived at the Cape almost a century before Maclear with the object of measuring the arc of the meridian in the southern hemisphere, and in addition, of cataloguing the southern stars. Lacaille's results, which apparently demonstrated that the earth was pear-shaped (thereby defying Newtonian predictions), aroused serious misgivings. One of his most emphatic critics was Colonel George Everest, the famous surveyor, who seeing the Cape on his way to India in 1820 suggested that "Lacaille's discrepant result lay in the gravitational attraction of Table Mountain".[44] Everest therefore recommended a remeasurement of Lacaille's arc, and Maclear in his letter of appointment was required to establish the best places for doing so.

In 1837 Maclear and his two assistants, Charles Piazzi Smyth and William Mann, began their task. After placing his zenith sector on the site of Lacaille's original observatory in Strand Street, he shifted it to the bottom of Platteklip Gorge in order to check the extent to which the mountain had affected the abbé's findings. The survey, completed in 1847, vindicated Lacaille's astronomy. "His anomalous result," as Professor Brian Warner notes, "was indeed owing to non-allowance for the gravitational attractions of mountains . . . [However] as it happened . . . Table Mountain lies to one side of the arc and has less influence."[45]

Having established Lacaille's reliability, Maclear built on that proof, finding that he could extend the trigonometrical survey he had done in Cape Town to Cape Agulhas. The stone cairn erected in December 1844 on the highest point of the mountain, 1 087 m above sea level, was part of that extension.

The floral abundance on the mountain has made it the lodestar of botanists since 1624, when Justus Heurnius stopped in Table Bay on his way to the East and collected plants on the mountain slopes, as undoubtedly did the great Paul Hermann and his pupil Heinrich Oldenland in their search for unusual Cape plants. Linnaeus's student, Sparrman, marvelled at the "rich harvest of the rarest and most beautiful plants".[46] "At hazard of my life," Thunberg – later professor of botany at Uppsala University – wrote, "I got for the first and last time the blue Disa longicornis, which is as beautiful as it is singular in its form."[47] "This last plant," he added, "grew in one spot only on a steep rock . . . in order to come at it . . . I was obliged to get upon the shoulders of M. Sonnerat when with a long stick I beat down five of these plants."

In 1773, Thunberg and Francis Masson, newly arrived from England to collect plants for Kew Gardens, "made an excursion on foot round the mountains situate between the Cape and False Bay".[48] Thunberg was particularly interested in those plants which the local inhabitants used for medicinal purposes, given the expense of importing European drugs. "They were very loud here in the praises of the *Osmites camphorina* . . . This plant has certainly great merit on account of its camphorated principle . . . The plant itself is sometimes used externally, applied in bags to inflammations and on the stomach in cholics . . . The genuine species, or the *Osmites camphorina*, which is the very best, I found growing on the top of Table Mountain only."[49]

Fortunately, the depredations of the woodcutters had little effect on these smaller plants. In 1810 Burchell botanized on the mountain with a continual sense of wonder. Walking around Lion's Head, "all that I had pictured for myself was far surpassed by what I saw in this day's walk! At every step a different plant appeared . . . I collected in four hours and a half, 105 distinct species of plants."[50]

Whether for plant collecting, exercising or indulging in man's fascination with a view seen from a height, climbing constitutes the most agreeable aspect of the mountain's human history. From the time that Antonio de Saldanha climbed Table Mountain to establish his bearings and in doing so "saw the end of the Cape and the sea that lies beyond it",[51] the ascent was to become an obligatory activity for healthy notables.

Thus Governor Simon van der Stel went up "with most of the folk of standing" (including the first white women to do so, preceding Lady Anne Barnard by a century). Having reached the top, "shots were fired and a large fire lit, and sometimes trumpets were gaily sounded and drums beaten, so that the wild beasts that might have been around there were chased away", Valentyn recorded.[52] In fact, the most dangerous animals would probably have been stone-rolling baboons.

Platteklip Gorge (Peter Mundy's "monstrous clefte . . . wondrous steepie"[53]) was the usual route; at the top, many exhausted climbers recorded the feat by painting, or scratching, their names on the rocks. The astronomers Maclear and Sir John Herschel, however, went up from the Constantia side, riding part of the way.[54]

Generally the ascent was begun either by moonlight, or in the cool of a summer's morning, at about 2 or 3 a.m., in order to watch the sun rise. Even the less distinguished climbers took a retinue of slaves (and later porters) to carry the provisions for breakfast and lunch, and to make coffee from the good mountain water. Both the Rutherfoord family and Louisa Ross, having dined well alfresco, took a nap in the shade, the latter under "two curious slabs of rock resting upon each other . . . nicknamed the 'Hen and Chicken' ".[55]

For some the climb was simply "a wearisome journey",[56] not to be repeated. "I have never spoken with anyone," Valentyn reported, "who had once gone up there and had a desire to go up again."[57] Thunberg, however, went up fifteen times in three years. It was agreed that the descent, though quicker,[58] was "more troublesome" even than the ascent, "requiring greater precautions to prevent our slipping and bruising ourselves amongst the stones."[59] Semple and Burchell ruined a pair of shoes each; while Thunberg's French friend M Sonnerat wore through three pairs of "thin French pumps" in one day.[60]

But most climbers would have agreed with Fitzherbert that "the prospect (being once up) fully compensates the labour".[61] Over two centuries later, Emma Rutherfoord enthused with a schoolgirl's hyperbole, "it was so wild, so free, so majestic, so worth the trouble and fatigue"[62] – though she too, did not want to repeat the experience.

Robert Semple's description of the view from the summit is lyrical in its symmetry, resonating still for the modern reader. "Beneath us lay the town with its gardens, its terraces and white, flat-roofed houses; the Table Bay with a surface smooth and unruffled by the slightest breeze, bearing on its glassy bosom numerous vessels of every nation, riding together peaceably at anchor. On the opposite side of the Bay, the high hills, somewhat inland, presented their tops covered with snow, and continuing to move ourselves slowly round, we observed in succession, once more, the dark mountains of Hottentots Holland; the flat, sandy space between the Table and False Bays; the False Bay opening outwards to the Indian Ocean; the mountainous south-eastern peninsula of Africa, on whose highest pinnacle we seemed to stand, stretching from north-west to south-east, and presenting a broken scene of mountains, precipices and chasms. Beyond it appeared the Indian Ocean and round again, nothing but sea, sea, sea till we once more came to the Lion Hills, the town with its white houses, Table Bay with its numerous vessels, and a little farther out, Roben Island [sic], in the mouth of the Bay."[63]

Up to 1890 most climbers, however competent, went up Platteklip and a few other uncomplicated routes. The establishment of the Mountain Club of South Africa in 1891 inaugurated an era of skilled climbing, using sophisticated equipment[64]; and since that date some 400 routes have been pioneered, at least 90 of which are on the front face. In addition, members of the Mountain Club may be regarded as the first custodians of the mountains, having from the outset observed and promoted a high code of outdoor ethics. The opening in 1929 of an aerial cableway on that "stupendous precipice"[65] has made it possible, however, to enjoy the grandeur without the physical effort or dangers of climbing.

For Table Mountain has long been a place of death. Burchell mentions an officer of the Cape garrison who, lost in the thick cloud which covers the mountain so rapidly in a southeaster, "at last fell over a deep precipice and was dashed to pieces".[66] In the 1860s Louisa Ross heard of "over five deaths" said to have been caused by people venturing too close to the edge of a sheer drop.[67] Such tragedies have been multiplied in modern times as growing numbers of people challenge the mountain.

Yet, on balance, the credo of General Smuts – and also that of his political adversary Cecil John Rhodes – dominates the human history of Table Mountain; the belief that the mountain is, above all else, a refuge and a place of great joy. "When we reach the mountain summits," Smuts said when unveiling the Mountain Club War Memorial at Maclear's Beacon in 1923, "we leave behind us all the things that weigh heavily down below on our body and our spirit. We leave behind all sense of weakness and depression; we feel a new freedom, a great exhilaration, an exaltation of the body no less than of the spirit."[68]

Human use of the mountain

JOSE BURMAN

Whether you approach Cape Town by land, sea or air, the outstanding feature of the landscape is Table Mountain. Thus it is; thus it has been for time immemorial. The mountain massif rises straight from sea level, its cliffs towering protectively above the puny man-made structures of the city. It renders Cape Town unique, for where else do you find a city whose heart is a playground furnished by nature?

Long before the first Westerners reached the Cape the little brown San would have been hunting buck in the kloofs of the mountain, and the Khoikhoi would have been using the lower slopes to graze their long-horned cattle and fat-tailed sheep. They made this possible by setting fire to the inflammable fynbos, for they knew it would be succeeded by succulent grass – a custom which the Dutch settlers continued when they occupied the Cape.

From the time passing mariners first saw the mountain, it drew them as a magnet. Not content merely to gaze at it, men imagined the glory of the summit; and the boldest among them climbed the mountain.

Antonio de Saldanha was the first man recorded as reaching the summit, but he cannot be included among the climbers for pleasure, since he ascended it in 1503 not for love but for information: to see whether his ship had passed the Cape of Storms or not. This was the first direct use of the mountain itself by Europeans.

The first real climber of Table Mountain appears to have been Peter Mundy, whom you met in Chapter 1. A passing English sailor, he climbed Table Mountain in May 1634 with two companions. They ascended by the same route as Saldanha – the great gash in the centre of the Table which is Platteklip Gorge, "the rocks on each side upright like monstrous walls . . . and had some rain mingled with snow at our going up".

Mundy was only the first of a long line of climbing adventurers, some more distinguished than others. In April 1655 there was Gysbert Heeck, who thought nothing of climbing the mountain while fasting, but arrived on the summit in an exhausted condition. Ten years later came Wouter Schouten and companions, refreshing themselves from a bottle of arrack instead of water, and arriving on top in a pleasant alcoholic daze. Then there was Nicolaus de Graaff, whose heartfelt words of thanks open Chapter 1 of this book and who returned 39 years later, in March 1679, to be the first man to sleep alone on the summit of the mountain: "But little sleep could I get, when I thought of the height of the hill, and the dangerous solitude of the same, where dwelt so many man-eating lions and evil tigers . . ." All he heard was the croaking of frogs.

The dapper French traveller François le Vaillant in the 1780s spent a night on the mountain and shot and roasted a vulture for his evening meal, but "even my dogs, after having smelt of it, turned away without touching it . . ."

July 1797 found the courageous Lady Anne Barnard climbing the mountain, dressed in her husband's trousers, which she used to good effect, sliding down from rock to rock. An unusual ascent of the mountain was that of James Holman, a blinded ex-naval officer who rode up Table Mountain on horseback in 1829.

Blockhouses

Apart from climbing it, from early times the inhabitants of the little settlement put Table Mountain to use in defending the Colony.

When Stavorinus visited the Cape its defences consisted of the Castle, the Sea Lines (connecting the Castle with Fort de Knokke), and the Mauritius Battery at the foot of Lion's Rump. In 1781 the French strengthened these fortifications by extending them up the slope (the French Lines). None of this hotchpotch of fortifications was tested when the British took the Cape in 1795, for the Dutch surrendered whilst the British were still at Wynberg.

General Sir James Craig, who now took over the defence of the Cape, was not impressed by the fortifications, as he had been preparing to bypass them on the mountain slope above the French Lines. He decided to build batteries and blockhouses closing this gap. First he built a battery and redoubt at the end of the French Lines, which he named York Redoubt. This has long since disappeared. Then, in 1796, he built a road across the slope to connect with the blockhouses he intended constructing. It is generally believed that Craig built three blockhouses on the mountain during his term of office, the Prince of Wales', King's and Queen's, before he handed over his command to Lord Macartney in 1797; however, a close perusal of the records does not support this.

From his secret report to Macartney it appears that Craig was busy on two fortifications (the Prince of Wales' and the King's Blockhouses). But even these were properly completed only after he left. In any event, by 1802 there were three blockhouses which had been built by the British.

The remains of the Prince of Wales' Blockhouse are reached from De Waal Drive. About 200 m south of the turn-off to the King's Blockhouse (a gravel road) is a little gate from which a path leads along the slope to the site of the blockhouse, overlooking Groote Schuur Hospital. Originally the blockhouse was a tower 9 m in diameter with walls 2 m thick. There were two storeys to the tower, and on the upper floor were three cannon. The blockhouse was demolished in 1826 and all that remains is a ruined building and a stone terrace on which stands a gun of 1782 vintage.

Queen's Blockhouse. (*Photo: Jose Burman*)

The King's Blockhouse is on the highest point of a ridge, just below the rocks of Devil's Peak, and is reached by a dirt road that branches off De Waal Drive before the Groote Schuur Hospital. It is a square tower in good condition, with gun embrasures at the top; it is a well-constructed stone fortification, not the makeshift building described by Craig. After the army had stopped using the King's Blockhouse it served for a time as a barracks for convicts. Today it is a national monument.

The Queen's Blockhouse, which figures on the 1802 map, fell into almost total disrepair, and was lost in a forest of wattle until rediscovered by the Cape Town Historical Society in the 1960s; it was left as a stabilised ruin, and stands just below the parking area at the end of Tafelberg Road. On the terrace adjoining the blockhouse are some 1755 cannon.

The final irony is that, having thus strengthened the defences, the British themselves had to retake the Cape in 1806. Knowing the strength of the blockhouse position, the British avoided it, and attacked along the edge of the bay. Since then there has been no further attack on the Cape, and the mountain defences were allowed to fall into disuse from 1827.

Water

The help which Table Mountain provided man in the matter of defence pales into insignificance when compared to the importance of Table Mountain's waters.

It was the water cascading down the mountain which made Table Bay the Aguada de Saldanha (Watering-place of Saldanha), and it was the same water which prompted the Dutch to set up a refreshment station there in 1652. Five streams flowed off Table Mountain: to the north the Varsche River; to the east the Liesbeek; south-east ran the Diep and Spaanschemat Rivers; and the Disa River emerged from Orange Kloof and ran south-west down the Hout Bay Valley. All these streams played a part in the expansion of the colony, but the true Aguada de Saldanha was the Varsche River, which ran down Platteklip Gorge to the seashore, where sailors filled their water-casks. It provided water for the entire settlement – drinking water, irrigation channels to the new garden, water for passing ships, and water to fill the moat which surrounded the Castle.

Nor was this the end of water requirements. The Dutch could never resist the idea of a "gracht" – the canal of their beloved homeland. Soon there was a Heerengracht, a Buitengracht and a Keisersgracht, crossed by little bridges, all fed by mountain water. During the 18th century the grachts began falling into bad odour – literally, for lazy slaves took to dumping night soil in the canals instead of in the sea. When the British took over they decided that instead of the people of the Cape having to go to the mountain for water, the mountain water would be brought to them. Water-houses were built, iron pipes were laid, and fountains were set up, from which water could be drawn.

The grachts were closed, and springs pressed into use, but Cape Town experienced continual water problems as the population grew. The water-mills which had been set up in Platteklip Gorge were closed down and two reservoirs (Nos 1 and 2) were built respectively in 1849 and 1856. But as the city grew so did the water problem. By 1872 the consumption had reached 400 000 gals (1,82 Ml) per day and the storage reservoirs were quite unable to cope with this. In 1877 work started on a large reservoir, the Molteno, which was completed in 1881. It was located at the foot of the mountain, partially excavated out of the slope, with embankments raised on three sides. As a result, on 27 August 1882 the reservoir wall gave and a flood of water poured down onto the inhabitants of Cape Town.

The reconstruction took 4 years, and long before it was completed it had become obvious that the water from Platteklip would never suffice to keep the Molteno full. "Honest" John Gamble, the Colonial Hydraulic Engineer, proposed a tunnel through the mountain to tap the unused water in Disa Gorge, and in 1887 work started on the Woodhead Tunnel. This was a major undertaking for those days – to drive a tunnel through a mountain, and construct a pipeline many kilometres in length, with tall aqueducts to carry the pipes to Kloof Nek, and down the mountain to the Molteno Reservoir.

The Woodhead Tunnel takes the shortest route through the mountain, emerging in Slangolie Ravine. Most of the space in the tunnel is occupied by the two pipes it carries, respectively 30 cm and 38 cm in diameter. Unless you are a dwarf, you have to bow your head to avoid the roof, and in places you must crouch to pass under projections. The tunnel runs more or less horizontally through the mountain, and there is only one slight curve. But for this curve you would be able to see the end of the tunnel from the entrance.

Part of the tunnel was arched in stonework, but most of it is raw rock. In places you can still see pegs, driven into the rock, and abandoned by the miners a century ago. The view from the exit in Slangolie is breathtaking. By 1891 the pipeline and the Pipe Track (a path which accompanied the pipeline along the side of Table Mountain to Kloof Nek) were completed, and the Woodhead Tunnel was opened with fanfare by the Mayor of Cape Town, Sir John Woodhead.

Laying the original pipeline from the Woodhead Reservoir in 1888.

Once again the authorities had not thought far enough ahead: the Woodhead Tunnel would faithfully carry all water running down from Disa Gorge, but in the summer this was practically nothing; the tunnel would be wasted unless there was a reservoir on the mountain to feed the tunnel. So in 1890 work began on the Woodhead Reservoir.

This reservoir was built of masonry (Cape Town had learnt a lesson from the earthen walls of the Molteno), and was 252 m long, stretching right across Disa Gorge; it was 44 m high in the centre. Although the sandstone blocks were quarried on the mountain, it must have been an immense task. For one thing it needed a railway to carry the stone, and the engine had to be carried up the mountain.

Top of the Kasteelspoort aerial cableway.

The cable-car running up Kasteelspoort.

An aerial cableway was devised, running up Kasteelspoort. The horizontal line of the cableway was 1 600 m, and it rose a vertical height of nearly 700 m. The load carried on the cableway was restricted to 600 kg. The wire hauling-rope was 3 650 m long and weighed 2 700 kg. On the summit edge of Postern Buttress the cable gear was set up, but the engine was located at the foot of the cableway. The cable-car was an open skip, on which passengers could crouch amidst the equipment; in fourteen spans it rose up the mountain, the last span being 450 m long.

From the summit station the railway line was laid and the little steam train was used to tow supplies to the dam. A minor township came into being on the plateau at the head of Kasteelspoort including shops, post office and a savings bank, since the workmen lived on the mountain for several years.

The last stone of the Woodhead Reservoir was laid in 1897 by Sir John Woodhead. The reservoir had a capacity of 210 000 000 gals (955 Ml), but to the surprise of the authorities, they found that consumption had already overtaken the increased supply. This little problem, which is the key to the water supply, has come up with monotonous regularity in Cape Town.

Fortunately the Woodhead Reservoir lay across a catchment of 255 ha, capable of yielding 500 000 000 gals (2 270 Ml) annually. So when work started on the next reservoir in 1898, it was sited immediately above Woodhead. The new reservoir was opened in 1904 by the Governor, Sir Walter Hely-Hutchinson, after whom it was named. Its masonry wall is 528 m long, but its maximum depth is only 16 m. The reservoir has a capacity of 203 000 000 gals (920 Ml).

While Cape Town was struggling with the Woodhead and Hely-Hutchinson, Wynberg (then an independent municipality) decided to follow Cape Town's example and built its own reservoirs on the mountain. Two reservoirs were constructed on a tributary of the Disa River, which also drained into Orange Kloof. The Cape Town reservoirs faced west, while the Wynberg ones faced east.

Victoria (the upper one), completed in 1896, has a capacity of 28 000 000 gals (127,3 Ml); Alexandra, below it (completed 1903), has a capacity of 27 000 000 gals (122,7 Ml). In 1903 the Mayor of Wynberg reported proudly: "With the completion of the work on Table Mountain there need be no anxiety for a shortfall next season."

Unfortunately Wynberg had not learnt the lesson of supply and demand, which Cape Town was learning the hard way, so in 1907 it was forced to construct yet another reservoir, the De Villiers, below the other two. The De Villiers faces west, like Woodhead, and is poised on the edge of the mountain overlooking Orange Kloof; although its capacity is only 53 400 000 gals (242 Ml) the depth of the water is nearly 30 m in the centre.

Wynberg municipality had experimented, not very successfully, with an aerial cableway during the building of Victoria and Alexandra, and when it came to building De Villiers they opted for a Trolley Track. This had a set of lines laid on sleepers running up from just outside the present Kirstenbosch Gardens. There were two trolleys, one going down with ballast as the other ascended. The motive power was supplied by two horses walking around a drum at the top station, but they were soon replaced by machinery.

In 1905 Cape Town experienced its next shock when it was discovered that the water consumption had risen to over 2 500 000 gals (11 Ml) per day. The Table Mountain reservoirs could only provide 3 000 000 gals (13,6 Ml) per day, so Cape Town was once again in trouble. Several alternative schemes for additional dams on the mountain were considered, but the bulk of Table Mountain water was already being utilised. Cape Town reluctantly turned to more distant sources of water and developed the Steenbras Reservoir in the Hottentots Holland mountains.

The water from Woodhead and Hely-Hutchinson continued to be used, and all of it passed through the Woodhead Tunnel. Slangolie Ravine, where the tunnel emerged, does not mean "snake oil" as its spelling indicates. Its real name appears to be "Slang Gully", but to the Cornish miners who built the tunnel a gully was a "goolie". Slangolie was a bad choice, for the ravine surface was becoming more and more unstable and rockfall succeeded rockfall. The ravine began to encroach on the tunnel. In the 1950s it became clear that if the water was to be of further use to Cape Town, it would have to be rerouted. As a result work started on the Apostles Tunnel in 1960.

This tunnel has its exit in Wood Buttress, a couple of kilometres nearer to Kloof Nek than Slangolie. At 1 300 m long, the Apostles Tunnel was a much more ambitious undertaking, and ran into considerable problems – loose sand through which the tunnel had to be pushed a metre at a time, underground water, and liquid mud. Concreting was used and the worst area was bypassed – the tunnel runs 835 m on the level and then rises 198 m over the next 450 m of tunnel, before emerging in Disa Gorge.

Despite everything they holed through the mountain on 14 October 1961 and the tunnel has now been in use for over 30 years, doing yeoman service. The Woodhead Tunnel is no longer operational, but the waters of Table Mountain still help to allay the unquenchable thirst of the city.

Constructing the new Apostles Tunnel, which now carries the waters of the Table Mountain reservoirs through the mountain.

Mining

No sooner had the miners finished tunnelling through Table Mountain for water than other miners began tunnelling into the mountain in search of tangible wealth – gold. The fantastic gold strike in America and Australia, the River diggings and Lydenburg gold in the Transvaal had an unsettling effect on the community of Cape Town and many were the rumours of gold at the Cape.

The best gold story of the period is that of the Cape Town shopkeeper who exhibited a nugget of gold in his shop-window, with the information that it had been found on Table Mountain. This sparked off a mini-gold-rush that sweated up Platteklip. When they reached the Silverstream Cave, halfway up the mountain, they were cheered to find the shopkeeper ensconced therein, selling cool drinks and beer and such accessories as picks and shovels which diggers might have forgotten in their haste to take part in the rush. No gold was ever found by the diggers, and when they eventually returned to the cave they found the shopkeeper long gone.

The belief that there was gold in Table Mountain persisted, and in December 1887 an item in the *Cape Argus* stirred the town into "gold fever". It read: "The Lion's Head (Cape Town) Gold Mining Company (to be registered under the limited liability act) . . . is formed for the purpose of carrying on the work begun by the Lion's Head Gold Syndicate, who have been prospecting on the farm near Lion's Head for the last 18 months, having met with most encouraging results . . . They have made one shaft of 145 feet deep and struck a reef 16 feet thick widening as it descends, the quartz from which upon assay from a ton crushed at Wilkinson's Mill, Kloof Street, gave over 2 ozs . . . A second shaft, between two distinct walls, is now being pushed on and an assay of stuff at 30 feet gives 8 dwts to the ton. A third (a pit) has been made on a very powerful reef, a quarry in fact, which gave at the surface 8 dwts to the ton . . . Applications for shares will be received from 10 December . . ."

These high hopes seem to have been dashed by a report on 29 August 1887 that the members of the Lion's Head Gold Syndicate were furious, as they had sent 7,5 tons of stone to England to be treated by the chlorination method, apparently hoping for an encouraging report which they could use in the floatation. Instead, only 10 lbs had been assayed, and that without recourse to chlorination.

This seems to have killed the operation, for the Lion's Head syndicate disposed of the property in 1892, leaving a gaping shaft in the hillside which was not even properly fenced off. The City Council threatened action, but apparently never did anything about it, for in 1951, during a major fire on the mountain, two fire-fighters had a narrow escape from death when they nearly plunged into this shaft in the dark. After that the shaft was filled in, and nothing remains of the attempt to mine gold on Lion's Head.

The tin mine on Devil's Peak was a venture of quite a different character, since it operated for some years as a thriving mine. Tin is a metal we see so often that we never bother to think of its origin. The biggest producers of cassiterite (tinstone) are Malaya and Bolivia.

The *Cape Argus* of 16 January 1912 carried this intriguing report: "Tin is being fetched out of a shaft which, after a little tunnel straight into the hillside, goes 180 feet down into a highly mineralised zone. Out of this shaft the Vredehoek Tin Company are bringing ore which will yield 2.5% tin . . . The crushed ore is in the form of cassiterite (tin oxide), which contains about 70% pure tin . . . about 100 men altogether are engaged on this mine."

Unlike the gold mine, the tin mine has left considerable remains on the mountain. It lies above Vredehoek in a little ravine. The first visible evidence of the mining activities is a line of concrete structures resembling horse-troughs, each one about 50 cm long and 50 cm

The tin mine on Devil's Peak in operation, 1912. (*Photo: South African Library*)

deep. Set end to end they rise in a series of steps, so that each four wash-troughs (for this is what they were), stand on a slightly higher level than the preceding group. At the top of this chain of troughs stands a large concrete reservoir about 8 m long – empty but well preserved.

The hillside above the reservoir has been excavated into several terraces on which stood the rock-crushers. The crushed rock was fed into a stream of water flowing from the reservoir, and carried by the water down the line of wash-troughs; here the tinstone, having a higher specific gravity, separated out and settled, while the lighter rock was washed away.

Inside the mountain there are even more signs of mining. There are two entrances, a higher one which leads to a vertical shaft (so care must be exercised in following this tunnel), and a lower one that runs into the mountain. If you follow this one for about 100 m, in pitch darkness you will come to a chamber with a pool of water in it. This pool is the head of a flooded shaft, so the water is deep. The tunnel continues for another 100 m beyond the pool.

What caused mining activities to cease? Was it the outbreak of World War I? Did the expense rise too high? Did they run out of tinstone? Whatever it was, the mine closed and the machinery was removed. Only the concrete dams, the reservoirs, the troughs and the honeycombed mountainside are left to remind us of the forgotten tin mine of Devil's Peak.

Mountaineering

Mountaineering was an extension and a logical outcome of walking on the mountain. By the early 1890s no more than a score of routes up the Table Mountain massif were known: half of these were straightforward walks, and the rest involved some climbing. People who were used to the Alps, where a corps of guides provided leadership and safety, were at a loss on Table Mountain, where there were no guides; the way was open for serious accidents.

A typical example of what could happen was provided in June 1889. A party of three men climbed Table Mountain via Platteklip Gorge, the easiest route. On the way down they tried a different way, following an apparently easy path down Fountain Ravine. The way became steeper and steeper and they ended on a narrow ledge. One of the party overbalanced and all three fell over the edge. Miraculously none were killed, and one of them was able to scramble down the mountain for help. A rescue party arrived late that night, and next morning was able to rescue the helpless pair, who were too weak to open a bottle of brandy lowered to them during the night.

The outcome of this accident was the formation of the Mountain Club in 1891, one of whose aims was mountain rescue. In addition they recruited a group of guides. There were eight guides, all Malays. This was not surprising, for at a time when most people regarded a trip up Table Mountain as high adventure, the Cape Malays were burying their dead on the mountain, and some of their holy tombs were located there; as a result many Moslems were intimately acquainted with the mountain.

The main purpose of a mountain club is mountaineering, and this the club proceeded to encourage, and very successfully at that. Table Mountain became the cradle of mountaineering in South Africa. During the century that followed Table Mountain was climbed by tens of thousands of members of the public, mainly following easy routes. But most climbers of consequence belonged to the Mountain Club, and their exploits and routes are recorded in an annual journal.

From early days mountaineers began grading the difficulty of climbs. In England they were using terms such as "easy", "moderate", "difficult", "very difficult", "severe" and "very severe". In South Africa they graded the climbs by letters of the alphabet, starting with A, which was a walk, and rising through B, C, D, E, etc.

The story of mountaineering on Table Mountain is largely the story of brilliant individual climbers, who led groups and set the pattern of climbing. The man whose outstanding climbing career established South African mountaineering on a firm footing was George Travers-Jackson. A natural climber, an infant prodigy, he was accepted into the Mountain Club at the age of 14 years. Travers-Jackson had an active climbing career of over 70 years, but his activities were at a peak from 1895 to 1907.

During that period he took part in no less than 51 first ascents on Table Mountain, of which thirteen were of D standard, seventeen of E, three of F and one of G standard. To appreciate this feat it is necessary to bear in mind that climbing in South Africa was in its infancy, that the technique of climbing was still to be learnt, and that safety precautions were totally lacking. The only safety instruction at the time was: "If your route is dangerous take a rope to assist you should you have to retrace your steps." The use of a rope to safeguard a leader was unknown.

Climbing solo, Travers-Jackson notched up a dozen new routes, including four of E standard and one of F.

The truth was that Travers-Jackson enjoyed climbing alone, and was extremely good at it.

The Anglo-Boer War of 1899 to 1902 seems to have affected mountaineering detrimentally, and the year 1910 marked not only the coming into being of the Union of South Africa, but a turn in the tide of mountaineering, which had reached its nadir. The early mountaineers had climbed routes of up to D standard; Travers-Jackson added seventeen E routes. The new school of Cobern, Fraser and Londt raised the standard and concentrated on F climbs.

These three climbers were to dominate the period from 1911 to 1927 on Table Mountain. W T Cobern first sprang into prominence in 1911 with the opening of a sensational climb on Slangolie Buttress. During the next 6 years he led four new E climbs and thirteen F routes on the mountain. The conquest of these new routes was made possible not so much by any major change in climbing technique, as by a concentration of competent and daring climbers. In fact, reading of Cobern's rudimentary rope technique and complete lack of safety precautions is enough to make one's hair stand on end. One change in climbing equipment did assist them. That was the introduction of rubber-soled sand-shoes, which the group often used when tackling difficult rock. Needless to say, this innovation was bitterly resisted by members of the old school as "unsporting".

After an orgy of frontal routes in 1914 Cobern's party was broken up by World War I. Climbing went on, with Fraser taking the position previously held by Cobern. J H Fraser was a splendid leader and soon built up a reputation for safe climbing. He was probably the first major climber in South Africa to introduce proper safety methods of using the rope, and members of his party learnt how to protect a leader in case he fell. In 1919 Fraser left Table Mountain and began climbing country peaks.

At about the same time George Londt was released from the army and returned to climbing. Although both had climbed with Cobern, Fraser and Londt were quite different in character: Fraser was quiet, confident, sure of his own ability, a graceful and rhythmic climber; Londt, on the other hand, was short of stature – a pocket Hercules – and he was often needlessly sensitive. The difference in their outlook ensured that Londt would not join Fraser's party. Instead Londt collected a group of his own.

Londt was handicapped by his size and sometimes found it necessary to use his second man as a stepping-stone, jamming him into a crack and climbing over him. This shortcoming was more than outweighed by his agility and strength, and his genius for finding a route up a rock-face. Londt refused to accept overhangs as impossible barriers, and was years ahead of his time. In another respect, too, Londt was an innovator, for he insisted on the use of belay points for a rope, to safeguard both the leader and the party.

With a very strong party Londt began climbing F standard climbs both on Table Mountain and on country peaks. He also introduced the idea of difficult descents. Until his day it was the custom to climb a difficult route and then descend by an easier one, but now Londt began descending by some of the most difficult routes then known on the mountain. Londt was killed in a fall on Table Mountain in 1926 – controversial to the last, for nobody was able to establish exactly why he fell.

Londt's second on many of his best climbs was Frank Berrisford, and Londt's mantle would have descended on him, had he not retired from climbing soon after Londt's death. Frank's younger brother Bert Berrisford became one of the most important climbers of the 1930s, leading over 120 first ascents. Difficult climbing was no longer confined to Table Mountain, and Bert Berrisford did most of his climbing on country peaks, but he did open 22 new routes on Table Mountain.

Someone who played an even more significant part in the Table Mountain climbing story, however, was Brian Cameron. During the period 1935 to 1937 he opened seven new climbs, three of which were graded G.

Though mountaineering in South Africa had thus reached G standard rock work, it is clear that this was due to the ability of the climbers alone, for technique had lagged behind, and safety precautions were quite inadequate in the event of a leader-fall.

The rules of South African rock-climbing in the 1930s were relatively simple: only one man at a time climbed, and he was the only one roped; no bags were left tied to the rope while a member of the party climbed; wherever possible the leader, when climbing, passed his rope over projections on the face, rocks or trees; when the leader brought up the other members of the party he kept the rope round a belay, not allowing slack to develop.

All these rules were sound, but it was in the safeguarding of a leader that the chief weakness lay: the rule was that the leader must not fall, for the best the second man could do was to pass the rope round a belay point and hold it firmly if the leader fell. Yet leaders did fall, as Brian Cameron's party discovered in 1937 when Alan Sluman fell to his death on Slangolie Buttress. On this occasion the second man did not have the rope round a rock belay (which would undoubtedly have snapped the rope), but tried to halt the running rope with his hands, suffering severe lacerations before the rope was torn

Girdle climbing at Kloof Corner. (*Photo: Jose Burman*)

A Tour de Force on Super Roof pitch.

from his grasp. This pointed the way to the answer – the dynamic belay, but that would only be introduced after World War II.

The principle of the dynamic belay was that, instead of passing the rope round a rock or tree as before, the rope would run round the body of the belayer, who was himself anchored to the cliff. The belayer's body would take the friction of the running rope and eventually arrest the fall. To establish an anchor, pitons (metal pegs) were hammered into the cliff and the belayer was attached to the piton by a snap-link of metal called a karabiner. The success of the new belay was so dramatic that of ten leader-falls in the next 10 years there was not a single fatality.

A second revolutionary technique, introduced from overseas, that now became popular was the abseil – a method of roping down a vertical pitch and even over an overhang with speed and safety.

The third factor that changed climbing after World War II was the arrival of the nylon climbing-rope, which had considerable stretch in it and so did not break as easily as the old Alpine climbing-rope.

It is hardly surprising, therefore, that after World War II there was a resurgence of climbing on Table Mountain which produced a powerful group of climbers, for these were the first to master the new techniques and equipment. The new style of climbing meant that the leader did not usually lead every pitch, and often the party took it in turns to lead. It is accordingly more difficult to pick out the outstanding climbers of this era, as there was a large group of almost equal ability, all capable of leading the most difficult climbs.

Perhaps the most enterprising of them were L D Schaff, Barry Fletcher and Mike Mamacos. Schaff dominated the immediate post-war period, the other two arriving on the scene in 1949. Barry Fletcher played a major role in the opening of new routes between 1949 and 1963, having a hand in no less than fifteen new climbs on Table Mountain. "Touch and Go" (G), opened by Fletcher in 1961, remained for years the hardest climb on Table Mountain.

The crux pitch on this route was extremely strenuous and hair-raising; none of the party was particularly keen on leading, so they drew lots. Paul White lost, and disappeared round the corner under an overhang. As they believed it would not be possible to reverse the pitch they waited in some trepidation. They heard the sound of a busy piton-hammer, followed by a gasp: "Just in time, I couldn't have held on much longer." As it was, White was so tired that he could not manage the last couple of metres and came down, being pulled back to the ledge after dangling below the lip of the overhang. Richard Williams, the third member of the party, completed the pitch. The piton which White placed was to hold a succession of sensational leader-falls over the next 10 years.

Mike Mamacos did his first pioneering route with Schaff on a country peak, and much of his climbing was in the country. On Table Mountain he was responsible for some fine routes, however, one of the best being Cableway Crag (G). The main problem was a long traverse which goes through the overhangs and ends in a stance well out from the rock-face below. As the second man could not follow the 13 m crux hand-traverse, Mamacos finished solo up a pitch every bit as hard as the crux.

Grotto-Fountain-Cairn traverse ledge. *(Photo: Mike Scott)*

"Rusty" Baillie did some climbing with Mamacos, after which he led a succession of hair-raising routes on Table Mountain, and Keith Fletcher (Barry's brother and a schoolboy at the start) was another great climber of the 1960s.

Anyone surveying the mountain scene in 1965 might have believed that climbing had reached a peak, and that little more in the way of new routes could be opened. During the period 1891 to 1939 some 242 new routes had been pioneered on Table Mountain, of which 130 were E/F standard. During the following 25 years another 116 new routes had been added of which 104 were E/F standard. There was room for only difficult new routes on the mountain, and the search for new ways was becoming harder and harder. Nevertheless the 10 years from 1965 to 1974 were to see an explosion of climbing such as Table Mountain had never experienced.

The key to this was the new technique of "mechanical" climbing. All advances in climbing technique and equipment had been resisted by conservatives, and mechanical climbing was regarded by the great majority of climbers of the day with intense suspicion. P E Anson-Cook put it succinctly: "A distinction surely should be made between methods which ensure the safety of life and limb under physical effort, and methods which merely make the impossible possible."

The proponents of the mechanical school considered this distinction not valid. They argued that many an impossible climb was possible if the safety of the leader was assured, and that even as far back as the 1890s Travers-Jackson had used a rope from above to safeguard a leader; the new mechanical aids would achieve leader safety and open up new fields.

The question whether these aids should be used for climbing on – pitons used as footholds, and étriers (ladders) to stand on – was much more delicate, and here only the out-and-out gymnasts were in favour. It would require a new school of climbers to overcome these objections – daring young men who had not been set in the mould of the past; who were keen to do free climbing, but would not disdain any help mechanics might provide in their conquest of more and more difficult climbs.

Michael Scott dominates the next quarter-century.

Nobody, not even the redoubtable Travers-Jackson, opened more climbs on Table Mountain, and nobody has produced so many climbs of a consistently high standard. During a climbing career that started in 1965 Michael Scott has opened more than sixty routes on Table Mountain, of which all but six were of F standard or higher, eight were of G standard, and nine were partly mechanical. He is the only climber who could claim in 1976 that he had climbed every route then known on Table Mountain. In 1988 Michael Scott was still opening routes of F3 standard.

Another important contribution which Scott made to mountaineering was his encouragement of other climbers. At one time or another Scott must have climbed new routes with nearly thirty different climbers, of whom thirteen became leaders of new climbs in their own right.

During the 10 years from 1965 to 1974 some 111 new routes were opened on Table Mountain, all of them of E standard or above. This tremendous burst of climbing may be ascribed to a combination of factors. These include the development of the latest modern equipment, the vast experience gained and passed on from generation to generation of climbing, and the knowledge that a safe retreat is assured by mechanical means. By the 1970s pitons were no longer fashionable; instead there were "nuts", which could be fitted into many places; "panic buttons" wedged into slots, formerly good only for pitons; "super-hexagons" to use in wedge-shaped cracks; "acorns" with hard metal ridges to provide a grip; and "beatons", soft lumps of malleable metal to hammer into even the slightest grips.

The number of top climbers of this period is far too great to enumerate, but one tendency emerged: the top climbers were becoming younger and younger, and schoolboys were in the forefront of climbing on Table Mountain. In the 1960s there were Keith Fletcher and Jan Goedknegt; the 1970s produced Chris Lomax and Greg Lacey; the 1980s brought to the fore Andrew de Klerk, and the 1990s Jeremy Colenso, Jonathan Fischer, Jeremy Wilse-Sampson, Adam Ross and Mike Hyslop. These Young Lions were now doing free climbs of H standard.

The ever-rising standard of climbs is due partly to the fact that leader-falls are no longer as vital as they were; with the artificial aids available, falls are becoming everyday matters, usually without serious consequences. This has encouraged the self-confidence of the leaders, and they have pressed to the edge of safety and sometimes beyond.

Another new development has been the increasing popularity of solo climbing, and this on some of the most difficult G routes. What motivates the young solo climber? Andrew de Klerk discloses his feelings in his description of Angel Dust 24, an H which he climbed solo: "Reaching my limit as I hang hard onto the tiny thing holding me from eternity . . . moving up, making that delicate balance transfer and laying away . . . fingers locked solid with fear against the hold. Need to be sure as I reach through to the rail, body tense, tenuous, and chest burning with the drug of life, and – I have it! Charge release surges me across the rail to the rest. Breather, shake-out and continue. Dancing on a high-wire up the wall, climbing lost in a mechanical desert surrounded by the stormy tumult of my mind . . . I pull through and sit here on the top, shaking . . . Cheating chance and playing chess; beating the opponent one move from checkmate. This is it – the voyage of self-discovery."

Solo climbing has become increasingly popular.

And so the wheel has come full circle: from Travers-Jackson opening 51 new routes on the mountain to Michael Scott with sixty top-class climbs; from boy wonder Travers-Jackson to schoolboy prodigies Jeremy Colenso and Jeremy Wilse-Sampson; from Travers-Jackson opening new climbs solo, through a century of ever-increasing climbing expertise, of teamwork, of safeguarding a leader by dynamic belays and a bewildering range of mechanical appliances, to Andrew de Klerk and Jeremy Colenso climbing solo on some of the most difficult rock in the world.

A puzzled onlooker might well be pardoned for asking "quo vadis?"

Caving

Everyone associates Table Mountain with mountaineering. Far less well known is the fact that Table Mountain is the happy hunting-ground of the caver. In fact, Table Mountain was the nursery of potholing in South Africa.

When a South African thinks of caves he visualises the Cango Caves, with their wealth of calcite formations and the thousands of metres of passages and chambers. A cave system such as Cango can only exist in limestone, which is easily penetrated and enlarged by slow-moving, weakly acidified water, into which carbon dioxide has been dissolved.

This does not mean, of course, that caves cannot exist in other rocks, but they are usually small. Exceptions are the caves of Table Mountain, which, though not among the biggest sandstone caves in the world, are quite respectable in size.

The caves occur in the ridge overlooking Orange Kloof, on the lower plateau of Table Mountain, and are vertical pothole systems. One of them has been known for over a century under the name of Table Mountain Cave or Wynberg Cave, for it lies opposite the reservoirs built by the Wynberg municipality. Travers-Jackson used to take parties down Wynberg Cave in the 1890s, and a partial survey of the cave was made in 1909. Until 1950 Wynberg Cave and a large open cavern known as Bats Cave were the only ones known, and no real cave exploration had been done on Table Mountain.

The only other sandstone caves in the Cape Peninsula are the large group of caves in the mountain above Kalk Bay, into which I was taken by some of the St James boys in 1947. I was not impressed. But in 1951 one of the same fellows offered to show me Bats Cave, which we approached over some very jumbled rocks. The cave, as my friend knew it, was a large vault about 12 m high and as broad; it was easy of access, and bats were always to be seen in it.

For a time I was content to stare at the bats' wheeling flight and listen to their outraged squeaks. Then it occurred to me that the number of bats was diminishing though no bats were leaving the cave. I watched more closely and saw a couple dart into an opening among the rocks; they did not come out again. I poked my head into the opening and found myself staring into the darkness of a lower cave.

My friend had no knowledge of the lower cave, nor was he anxious to find out, but my curiosity was stirred and I resolved to explore that cave – thereby starting a train of events that led to the discovery of most of the Table Mountain caves, and the formation of the SA Spelaeological Association (SASA).

At first I had great difficulty in persuading anyone to join me, and the early exploration of Bats Cave was done alone, using my climbing-rope and a hand-held torch – a somewhat risky state of affairs. I didn't seem to get anywhere, for there was usually insufficient time, light or rope to make further descents. Eventually I persuaded one or two other climbers to come down with me, but once was enough for most of them; the only one who was bitten by the caving bug was Adrian Bogers, and he became my constant companion for the next year as we explored Table Mountain looking for caves.

In 1952 Adrian, Jerry Wannenberg and I found a well-lit cavern, its floor filled with huge tumbled rocks. When we climbed down into it, it looked as though a giant hand had tossed the rocks around in careless abandon, and we named it Giant's Workshop. We were not particularly impressed with the cave, which shows we were very green!

Instead we continued our surface exploration, and not far from Giant's Workshop we struck gold. The entrance was a small hole which gave access to a vertical pothole. As a result we built our first rope-ladder and used this to explore the depths of the cavern, which we named Climbers' Cave.

In 1953 we turned our attention to Wynberg Cave, which had two entrances, one a vertical leading to Pluto's Hall; but we had noted a hole about 80 m from the Wynberg entrance into which we descended. This gave access to a vertical cave, which we believed was Wynberg. So the next weekend we came back with reinforcements. One party entered Wynberg Cave itself, while another climbed down our new hole and made contact with them – it was all part of Wynberg. Next we discovered two further holes in the mountainside, and found that they connected with Wynberg by long ladder pitches. So by the end of 1953 Wynberg's length had been extended to an estimated 300 m and there were five entrances.

I passed through the opening and it was indeed a find. A high passage led downward at a gentle angle into the darkness; it became a rift, necessitating wall-walking, and then descended from ledge to ledge in a rough stairway. A few weeks later we were back with a strong party, fully equipped. We found a maze of passages – enough for a Yorkshire potholer who visited us a couple of years later, to say blissfully: "Ee lad, soom cave."

Giant's Workshop, however, was nothing compared with the surprises which the little Smugglers' Cave gave us. An unobtrusive entrance among rocks on the lower plateau gave access to a short series of tunnels opening into little caves with windows onto the main cavern – a children's paradise in which to play smugglers. But in a corner 6 m below the entrance there is a small shaft. We tackled this shaft in 1954 and it led to a small chamber, the entrance of which had to be excavated before we could crawl through. The chamber proved disappointing – a dead end.

As we were leaving it I put my hand under a rock and found a hole beyond which I could feel nothing. Lightly belayed, I wriggled under the rock, feet first, found no bottom, so planted my feet on the opposite wall and

Caving inside Table Mountain's Climbers' Cave. (*Photo: Jose Burman*)

Exploring Smugglers' Cave. (*Photo: Jose Burman*)

By now our weekly cave-hunting trips had yielded a number of smaller caves such as North-west Passage, Hangman's Cave, the Powder Room, the Magistrate's Court, Adrian's Well and Smugglers' Cave.

In 1954 the arrival in Cape Town of a keen caver from England, Don Tritton, precipitated the formation of a caving club – the SA Spelaeological Association. Don became the first secretary and I the first chairman. We started showing the new members some of our finds.

One of our earliest meets was Giant's Workshop, which I reckoned was absolutely safe for beginners. There were 29 in the party and I turned them loose in groups of five to explore the cave. When the party reassembled there was one group short. Nobody had left the cavern, so we began searching in case they were trapped somewhere. Suddenly there was a shout and one of the missing men emerged in the far corner. When I joined him I could see no opening (it was hidden under a rock) but the excited chatter made it clear they had found something.

swung into the hole. To my horror I found myself clinging like a bat to the ceiling of a 6 m deep cave. My exit was jet-propelled. This was the entrance to the main Smugglers' system that kept us busy for the following year.

The cave I had glimpsed yielded two holes in one corner through which you could crawl; from there a ladder took you down to the next major cave – Piccadilly (the underground begins here). Indeed it did! A long ladder pitch brought you down to Snark's Lair. We took the press in with us on that trip and there were some hilarious pictures. We discovered more systems off Snark's Lair and spent months checking them out. One trip started at 20:00 on Saturday, laddering the cave; the day shift arrived at 10:00 on Sunday and took over. The ladders were removed on Sunday evening.

For several years SASA's caving was concentrated on Table Mountain; then interest shifted as SASA began the survey of the Cango Caves and a number of caverns were discovered in the limestone near Bredasdorp. Then came expeditions to the Transvaal, South-West Africa (now Namibia) and Rhodesia (now Zimbabwe).

SASA has now been functioning for 40 years and is more active than ever, but far more experienced than when it was formed. Regular caving meets are still held in Table Mountain caves, which the Association regards with affection, for that is where it cut its teeth.

The cableway

Until 65 years ago you had to be fit and in good training to enjoy the pleasures of Table Mountain. The quest for an easier means of access to the summit started a long time ago. In the 1870s, shortly after the introduction of railways at the Cape, Dr Van Oppel and Capt Van der Veen, two residents of Cape Town, began agitating for a rack-railway up Table Mountain – a train drawn up the slope by a locomotive running on cog-wheels which fitted into slots in the line, a system popular in Europe.

Nothing came of this agitation, but the Kasteelspoort Cableway revived interest, though the Anglo-Boer War and the postwar depression killed any likelihood of putting it into execution. In 1910 Cape Town became the legislative capital of the new Union of South Africa and in 1912 the Cape Town City Council invited a Swiss expert, Mr H M Peter of Zürich, to submit a report, which Peter produced in 1913.

Two alternatives were considered: the first was a rack-railway from Kloof Nek, up and over Kloof Corner to the Central Table via Fountain Ledge. This line would have been quite difficult, for it would have ascended the open slopes from Kloof Nek to the first rock-band and then by way of various cuttings and tunnels it would have reached Fountain Ledge, whence it would have run via Fountain Ravine to the top of Platteklip Gorge. This would obviously have disfigured the mountain, though Peter concentrated on the exceptional views that the passengers would enjoy on the ascent.

However, Peter favoured the second alternative, namely a funicular railway from Oranjezicht up Platteklip Gorge, with a middle station roughly where Tafelberg Road now runs; here the passengers would change trains for the steeper section of the route to the Central Table, where the upper station was to be located.

In August 1913 a poll was taken and a two-thirds majority of Cape Town's citizens approved an expenditure of £100 000 on a funicular. Work began at once, to be halted by the outbreak of World War I in the following year. It was not until 1925 that the scheme resurfaced, and by then it had changed completely.

The proposer of the new scheme was Trygve Stromsoe, a Norwegian engineer who had settled in Cape Town in 1925. In 1926 he produced a scheme for a cableway, instead of either a rack-railway or a funicular. Sir Alfred Hennessy became interested in the scheme, and he involved Sir David Graaff and Sir Ernest Oppenheimer. A company was floated with the four originators on the board.

It was not until 1929 that the 10,5 ton track-rope, which had been imported from abroad, was hauled from the docks up to the lower cable station, and eventually drawn up the mountain. The opening day was 4 October 1929, and the ceremony was performed by the Mayor of Cape Town, the Revd A J S Lewis in the presence of some 200 guests.

The cableway was certainly a novelty, but its early days were beset with problems, mostly financial, for it opened during the worst depression to hit the world. During the early 1930s both Oppenheimer and Graaff withdrew their money, and in 1932 Stromsoe dropped out, but Hennessy soldiered on. The end of World War II brought major improvement. The population was growing, the number of tourists multiplied, and people were becoming more interested in the outdoors. From then onward the cableway's success was assured, for the pace accelerated steadily to reach the first million visitors in 1959. The second million took another 11 years, and the third million 5 years. Five million visitors were added during the next 17 years and the total now is in the vicinity of ten million visitors.

At present nearly 500 000 visitors use the cableway every year, a remarkably high figure considering that for one day in five the cableway is out of operation. This is

The old cable-car running up Table Mountain, 1940. *(Photo: South African Library)*

not because of mechanical failure, but because of wind. The siting of the cable was done carefully, for it is in line with the two prevailing winds, the south-easter and the north-wester. A cross-wind such as the north-easter (the berg wind) can cause problems and the cessation of operations at even low wind speeds. The south-easter, on the other hand, must achieve a considerable velocity before the cableway has to be closed down, because the cable-cars travel in the same direction as the wind.

The cableway is particularly proud of its safety record, for it has carried nearly 10 000 000 passengers without a single casualty, stretching over a period of 65 years. This is no doubt due to the rigid safety precautions adopted by the company. Every week the cable is subjected to a detailed inspection over its entire length. This is a job carried out by a senior millwright and crew, who perform their hazardous duties on the roof of the cable-car as it travels up and down the line, ensuring that the wheels run smoothly and that the cable is well greased. In addition periodic electro-magnetic tests are conducted by independent experts.

The plant was renewed in 1967 and the main cable was renewed in 1958 and again in 1974. The cable-cars

A modern version of the cable-car above Lion's Head as seen from Kloof Corner. *(Photo: Jose Burman)*

have also been changed. The first cars served for 29 years and carried 1 250 000 passengers in loads of twenty at a time. The second set of cars operated from 1958 to 1974, carrying 2 000 000 people in loads of 24 at a time. The present cars are lighter and can carry 28 people.

Although there have been no accidents on the cableway, power failure on a couple of occasions caused embarrassment. There was the occasion in 1979 when 200 tourists were stranded on the summit for several hours during a general power failure. Even more unpleasant was the occasion when 44 passengers were stuck in two cars in mid-air for 90 minutes during another power failure. This should never occur again, for in 1982 the company installed a Rolls-Royce generator to take care of any electrical hiccups.

The only trouble with the cableway is that it has proved too successful. During the school holidays and over Easter the pressure has proved so great that long queues of impatient holiday-makers form, anxious to visit the summit of Table Mountain. Nor is it possible to avoid the queues, for the cableway at that time is run at maximum capacity. The period of 8 minutes taken by the original cars has now been cut to 5 minutes. When there are queues the turn-around before the next trip has been cut to a minute. In this way ten trips can be handled with 280 passengers per hour. The hours are extended, so that the day starts early and finishes well after dark during the season – but the queues are inexhaustible, for this is the second-most popular tourist attraction in South Africa – exceeded only by the Kruger National Park.

Doubling the cableway would be enormously expensive, increasing the size of the cars would not be cost-effective, and neither would justify the increase in expense and the higher charges. A second cableway, from the city to Signal Hill, has been suggested. This has possibilities, for the public could use one or the other, or even both in succession if a linking bus were provided.

In any event, the future of the cableway is undecided, for the Cape Town Municipality, which has always had the option of taking it over, is at present discussing the pros and cons of expropriation. It could well be that by the end of the century the Table Mountain Cableway will have an entirely new look. Of one thing, however, we can be assured: the cableway will not be abandoned, for this is one of the greatest tourist facilities offered by Table Mountain, and indeed, by South Africa.

Geology and climate of Table Mountain

DR JOHN MOORE

The great front wall of Table Mountain and the impressive sweep of buttresses, ravines and amphitheatres that line its flanks have quite justifiably been described as one of earth's natural wonders. Moulded and carved out by the forces of nature, the rocks reveal an event-filled past that spans almost 1 000 million years in the history of our planet.

The Cape Peninsula mountain chain is made up of a remnant mass of ancient lithified or compacted sedimentary rocks belonging to the Table Mountain Group (TMG). Rock formations of the TMG are relatively widespread along the southern coastal belt of South Africa, from East London and Port Elizabeth in the east to Vanrhynsdorp and Nieuwoudtville in the west. They make up many of the prominent mountain ranges in this region, including the Baviaanskloofberge, the Outeniqua and Kouga mountains, the Swartberg and Langeberg, the Hex River mountains, the Hottentots Holland range and Franschhoek mountains, the Du Toit's Peak massif, the Koue Bokkeveld range and the Cedarberg. Similar rock formations are found in Argentina, on the Falkland Islands, across the Transantarctic mountains and along the east coast of Australia.

All these sediments are thought to have accumulated at more or less the same time along the coastline of an ancient supercontinent called Gondwanaland, that split apart some 150 million years ago to form the present-day configuration of continents in the Southern hemisphere (South America, Africa, Antarctica, Madagascar, India and Australia). The sediments are Palaeozoic in age (palaeo = ancient; zoo = life) and contain marine fossils that are correlated with those of the late Ordovician period in Europe, making them approximately 450 million years old.

The TMG in the southern Cape region is subdivided into several different rock formations based on the different types of sediments found within a vertical cross-section of the group. On the Cape Peninsula, however, only two of these formations – the underlying, or lower, Graafwater Formation and the overlying, or upper, and

The great front wall. (*Photo: Zelda Wahl*)

far more prominent Peninsula Formation – are present as major units. A tiny remnant of the glacial sediments belonging to the Pakhuis Formation can be found in a small area near Maclear's Beacon on the summit of Table Mountain. The greatest thickness of TMG sediments preserved on the Cape Peninsula is slightly less than 800 m and can be observed in a vertical profile from Tafelberg Road to Maclear's Beacon.

It is most probable that additional rock formations belonging to the upper parts of the TMG (the Cedarberg and Nardouw Formations) were originally deposited on the Peninsula Formation, doubling the

Massive sandstone formations. (*Photo: Paul Alberts*)

height of Table Mountain as we see it today. These sediments have long since been weathered, eroded and washed away into the surrounding oceans. Similarly, rocks of the TMG would once have covered the Cape Flats and False Bay, ultimately linking up with those of the Hottentots Holland range. They too have eroded away with time. The Peninsula mountain chain persists as a rather enigmatic remnant that has resisted the weathering processes still at work today.

Sedimentary rocks like those of the TMG are formed in processes that occur at or very near the earth's surface. These processes initially involve the weathering, breaking down and erosion of older, exposed rocks by the action of wind and water. This is followed by the transportation of the weathered material as boulders, gravel, sand grains and fine clay particles by the forces of gravity, generally in suspension in streams and rivers, and finally by the deposition or release of this suspended load in lakes, lagoons, estuaries, on beaches, or in deeper marine areas along the continental shelf. Here the particles accumulate as horizontal sedimentary layers called beds or strata.

Sediments are initially classified by the size and composition of the transported grains (clasts) that they contain. The coarsest clastic sediments are termed conglomerates, in which the size of the particles varies from that of large boulders down to small pebbles. These are followed in decreasing grain size by gravels, sands or sandstones, and silts or clays – the lattermost generally smaller than can be seen individually with the naked eye. In thick successions of sedimentary rocks like the TMG, it is not unusual to find a complete range of rocks representing all these different sizes of sediments.

A wide variety of minerals are found in clastic sedimentary rocks, although in general only a few common minerals make up the dominant constituents of any rock type. In the coarser sediments, the major components are minerals and small rock fragments that were hard and tough enough to survive the rigours of weathering and transportation without complete disintegration. Quartz is one such durable mineral and it predominates in the conglomerates, grits and sandstones of the TMG. Although of similar hardness, feldspar, the other common constituent of crustal rocks, possesses cleavage planes; natural weaknesses along which the mineral breaks up into small fragments. Feldspars react with water in the weathering cycle and break down to clay minerals such as illite, sericite and kaolinite. Almost all the sandstones of the TMG are quartz arenites – quartz-rich sandstones with less than 10 per cent feldspar and rock fragments. Sandstones with larger amounts of feldspar are called arkoses, and a few occur in the lower Graafwater Formation.

Shales and siltstones, on the other hand, consist of fine clay-sized particles, generally formed by breakdown of less durable minerals. Typically, shales comprise about 60 per cent clay minerals, 30 per cent fine quartz and 10 per cent other minerals such as feldspar, carbonates (calcite or dolomite), organic matter and iron oxides. The purple siltstones and shales of the Graafwater Formation (and rarer bands in the Peninsula Formation) are made up of mixtures of the clay minerals illite, sericite and kaolinite together with quartz and the iron oxide mineral haematite, which gives the rock its characteristic purple colour. Small amounts of black manganese oxide (pyrolusite) are also present in the shales, and these oxides have, in places, migrated and recrystallised along joint planes and in fractures or faults.

The older basement

THE MALMESBURY GROUP

The sands and clays that make up the TMG sediments were derived from the weathering of considerably older rocks, such as the Namaqualand granites, in the interior of the Gondwana supercontinent. Similar older rocks also made up the floor or basement on which the TMG sediments were deposited. On the Cape Peninsula, this basement consists of two very different types of rock. The oldest rock types are metamorphic rocks – slates, quartzites and phyllites – of the Malmesbury Group. These are located beneath the TMG sediments in the northern part of the Peninsula and the area they cover includes Sea Point, Signal Hill, the city bowl, the lower mountain slopes beneath Devil's Peak, Groote Schuur Estate and the southern suburbs to Claremont. Rocks of the Malmesbury Group have also been intersected at shallow depths in boreholes drilled on the Cape Flats and make up much of the Swartland stretching from the Durbanville hills to Porterville, Piketberg and Saldanha. They are generally poorly outcropping rocks that weather easily to red clayey soils and can be best observed along the Sea Point-Green Point beachfront and in road cuttings and quarries along Signal Hill and on De Waal Drive. Besides being the oldest rocks (about 900 million years old) they are also the most disturbed or deformed and recrystallised. In most places the originally flat-lying sedimentary strata now dip at steep angles.

CAPE GRANITE

Over the rest of the Peninsula southwards, and underlying the major part of the Peninsula mountain chain, the basement consists of a massive, relatively coarse-grained igneous rock called Cape Granite. Granitic rocks form at considerable depths within the earth's crust, down to 80 km beneath the continental surface. Changes in temperature and pressure at these great depths cause rocks to start melting and a highly viscous, crystal-rich melt called a magma is formed. The magma rises through the earth's crust due to natural buoyancy, forcing its way upwards through the surrounding rocks. Sometimes these molten magmas reach the surface and erupt spectacularly as volcanoes, but more commonly they solidify and crystallise at depths of several kilometres below the surface. This happened in the case of Cape Granite, which formed by melting of the deeper parts of the earth's crust some 550 million years ago (well before the formation of the TMG sediments). It rose as a large molten mass (batholith), forcing its way through and into the older Malmesbury Group rocks, cooling and finally crystallising approximately 10 km below the earth's surface.

During crystallisation of the Cape Granite, large crystals of feldspar (phenocrysts) were the first to form, enclosed in a finer mass of grey quartz, black mica (biotite) and more feldspar. The large yellow-weathering feldspar phenocrysts are most distinctive, having a rectangular outline, breaking along flat, shiny cleavage planes, and reaching several centimetres in size in many places (for example, at Clifton, Bakoven and Llandudno beaches).

Outcrops of granite are most prominent on the western slopes of Lion's Head, along the western coastline from Clifton to Llandudno and on the upper slopes beneath the Twelve Apostles. Cape Granite outcrops are also found along the Catwalk at Fish Hoek, at Boulders, Seaforth and Miller's Point in the Simon's Town area, below Chapman's Peak and at the extreme tip of Cape Point. Weathered granite provides the fertile soils for the Constantia vineyards, and more intense weathering of the granite has resulted in the formation of kaolinite deposits in the Sun Valley and Noordhoek areas.

When the hot Cape Granite forced its way through the Malmesbury sediments, a transformation took place in the latter rocks in the vicinity of its contact with the granite. The intrusion took place at high temperatures (granites crystallise at temperatures between 700 °C and 800 °C) and it "baked" the adjacent sediments, forming recrystallised metamorphic rocks, called hornfelses. A spectacular "mixed zone" (migmatite), consisting of interfingering bands of granite and Malmesbury horn-

Granite boulders eroded by wave action at Llandudno. (*Photo: Paul Alberts*)

Interfingering bands of granite and hornfelses at Saunders' Rocks. (*Photo: Zelda Wahl*)

felses, can be seen at Saunders' Rocks on the Sea Point coast. This represents the contact zone between the molten granite magma and the older sediments into which the granite intruded and crystallised. This granite contact at Saunders' Rocks has a prominent place in the history of geological discovery and played a crucial role in arguments advanced by John Playfair in 1815 concerning the magmatic origin of igneous rocks. Charles Darwin made a particular point of examining and describing this locality on one of his famous voyages on HMS Beagle in the 1830s. The site is quite justifiably preserved as a national monument today.

Associated with the intrusion of the granite, hot solutions carrying ore minerals penetrated into the Malmesbury sediments close to the contact zone, where they cooled and crystallised as quartz veins. One of these veins on the lower slopes of Lion's Head above Tamboerskloof was mined for gold in the last century and others on the lower slopes of Devil's Peak contained tin mineralisation. In both instances the veins were small and impersistent and mining operations were short-lived.

After the intrusion of the Cape Granites, an extended period of almost 100 million years passed during which weathering and erosion of these basement rocks took place until they formed a wide flat coastal plain to the Gondwanaland supercontinent. It was on this coastal plain that the sediments of the TMG were deposited.

The Table Mountain group

THE GRAAFWATER FORMATION

Rocks of the Graafwater Formation immediately overlie the basement platform of Cape Granite and Malmesbury Group metamorphic rocks and the formation is, on average, about 70 m thick. The Graafwater Formation is most prominent from Lambert's Bay on the west coast inland to Citrusdal and takes its name from the village of Graafwater, west of Clanwilliam. The Graafwater Formation can be distinguished from the overlying Peninsula Formation by its reddish to purple colour. This colour is invariably associated with shale or siltstone layers, and signifies the presence of iron oxides such as haematite. Fine silvery flakes in the shale bands are concentrations of sericite mica. Sandstone bands in the Graafwater Formation have a distinctive yellow colour compared to the grey-white of the overlying Peninsula quartz arenites. This is due to small amounts of weathered feldspar in the sandstones.

Excellent exposures of the Graafwater Formation can be seen in road cuttings along Chapman's Peak Drive and beneath Devil's Peak on Tafelberg Road. The Graafwater sediments, being relatively soft shaly horizons, weather easily and are commonly covered with scree (angular unconsolidated rubble) from the overlying Peninsula sandstones. In other areas, outcrops of Graafwater rocks form small wind-eroded caves and overhangs beneath the harder sandstone units. This is typically seen on Lion's Head, along the front of Devil's Peak and beneath Kloof Buttress and on some of the Twelve Apostles. A characteristic feature of the Graafwater Formation is the horizontal layering or bedding, accentuated by the variations in colour between the purple mudstone bands and yellow sandstone bands.

Within certain horizontal layers of quartz arenite and purple mudstone, a secondary banding at an oblique angle called cross-bedding or cross-stratification can be observed. Cross-bedded sandstone units are extremely common throughout the sediments of the TMG and indicate that much of the sand and silt that make up the TMG found its way to its ultimate depositional site by being transported via a network of meandering stream channels. The angular cross-beds formed as the sand progressively filled these channels.

Another common and characteristic feature of the Graafwater mudstones and shales is the presence of ripple marks on the upper surfaces of the units or beds – so named after their similar appearance to water ripples. The ripple marks are best observed in plan (from above) and commonly have wavelengths of 5-10 cm. Some fine examples are preserved where the contour path crosses Silverstream Ravine on the face of Table Mountain. Ripples form wherever currents are present, particularly where there is tidal flow across mud-flats or in wide, shallow channels. Most ripples are slightly asymmetric, having a steeper advancing slope and a flatter trailing slope. They were in fact advancing slowly like miniature dunes within the tidal or stream current.

Besides cross-bedding and ripple marks, rocks of the Graafwater Formation show several other interesting but rarer structures. One feature is the presence of mud-cracks in the purple shale bands, with the cracks filled with coarser yellow sandstone material from overlying sandstone layers. The presence of the mud-cracks, which appear in plan in a polygonal "crazy paving" arrangement, indicates that the mud layers were exposed to the air, dried, shrunk somewhat and as a result cracked open, prior to the deposition of the overlying sands. This tells us that these fine muds were deposited in very shallow lagoon and tidal-flat environments that periodically became exposed to the sun and dried out.

Other features preserved at these horizons are the fine tracks, trails and burrows of primitive bottom-dwelling

Cape Granite that underlies the Graafwater Formation at Chapman's Peak. (*Photo: Paul Alberts*)

Graafwater beds being eroded by wave action. (*Photo: Zelda Wahl*)

marine creatures such as trilobites. Although their tracks and burrows have been preserved, no body fossils have so far been discovered within the Graafwater Formation and therefore the precise age of the formation has never been determined. The lack of fossil remains is partly due to the oxidising environment present on the tidal flats, which did not allow for the preservation of organic matter. Body fossils of trilobites and brachiopod and mollusc shells have been found higher up in the TMG sedimentary sequence in the Hex River and Bains Kloof mountains. It is these fossils that have established the age of the TMG in general at around 440 million years.

During the period when the sediments of the Graafwater Formation were deposited, the area covered today by the Peninsula mountain chain formed part of a vast mud-covered tidal flat cut by meandering tidal channels filled with sandy material. Various arthropod species inhabited this otherwise barren flat, leaving their tracks and burrows on the tide-rippled and desiccated surface.

THE PENINSULA FORMATION

A distinct change occurred in this environment at the start of the deposition of the overlying Peninsula Formation. The change is marked by the virtual disappearance of fine-grained sediments like shales and mudstones and the total dominance of coarser-grained sandstones or quartz arenites. The Peninsula Formation is not only very thick (700 m on Table Mountain, increasing to over 3 000 m in the Koue Bokkeveld and Hex River mountains) but also extremely uniform in appearance. It consists almost entirely of quartz arenites that commonly display large-scale cross-bedding features, and locally contain thin impersistent shale or mudstone bands and thin conglomerate lenses. The pebbles in the conglomerate bands are commonly of a massive milky-white vein quartz with rarer quartzite and dark shale/slate pebbles. Large cliff faces such as those on the front of Table Mountain, the Twelve Apostles and Fernwood Precipice overlooking Newlands Forest provide ideal exposures of the massive and impressive sandstone beds, many of which are several metres thick.

The individual strata or beds of quartz arenite in general become progressively more massive and thicker as one progresses higher up the mountain through the stratigraphic sequence. Using the front of Table Mountain as an example, the Peninsula Formation can be subdivided in a simplistic way into three somewhat indis-

The Peninsula Formation at Chapman's Peak. (*Photo: Paul Alberts*)

Peninsula formations exposed at low tide. (*Photo: Paul Alberts*)

tinct zones. The lowermost zone which stretches from the contour path to the foot of the lower buttresses comprises comparatively flat or planar beds, followed by a central section covering the lower buttresses characterised by large-scale, lens-shaped channel deposits. The uppermost section comprises very thick, tabular to lenticular sandstone bands making up the final impressive vertical cliff faces. The central and upper sections with their large-scale channel deposits were formed in an environment where powerful currents were active. The channel deposits typically contain a thin pebble conglomerate band at the base, a central mass of sandstone and a finer-grained upper veneer of siltstone or shale.

Most of these environments were harsh, high-energy areas that were not conducive to the preservation of fossil remains of the marine creatures living there. Rare arthropod tracks and burrows are observed in certain thin shale bands, indicating that a similar fauna inhabited the coastal region to that recorded in the Graafwater Formation. The continental margin of Gondwanaland must have slowly subsided over millions of years while the great thickness of Peninsula Formation sands accumulated.

THE PAKHUIS FORMATION

At the end of this period of sandstone accumulation, a major climatic change resulted in the advance of a vast ice-sheet from the Gondwana continent to the north, out over the coastal areas. Remnants of the sediments that formed during this glaciation are located on the top of Table Mountain in the Maclear's Beacon area. Elsewhere along the Peninsula mountain chain, only the Graafwater Formation and portions of the Peninsula Formation are preserved as representatives of the TMG. The glacial sediments take their name from Pakhuis Pass in the northern Cedarberg mountains.

Sediments derived from glaciers or ice-sheets have unique characteristics that make them relatively easy to distinguish from other rocks. Pebbles, sand grains and clay particles are caught up in the glaciers as they advance slowly under the force of gravity. Coarse pebbles lying at the base of the glacier are ground against one another and the solid underlying basement rocks, creating distinctive facets and striations (scratch marks) on the pebbles and forming a fine rock powder. At the end of its journey, either in the sea or at the foot of some mountain range, the moving ice-sheet breaks up and starts melting, depositing a mass of pebbles, sand particles and fine rock powder that accumulates as a poorly bedded mixed sediment, termed diamictite or tillite. Bands of tillite are present in a small area on the mountain summit near Maclear's Beacon, and faceted and striated pebbles can be found there that have weathered out of these beds.

Subsequent to the deposition of the glacial sediments of the Pakhuis Formation, shales and siltstones of the Cedarberg Formation and sandstones of the Nardouw Formation were deposited as the upper portions of the TMG. Sediments of these latter two formations are no longer preserved on the Peninsula, where it is presumed that they have been stripped away by more recent erosion. The flat top of Table Mountain is thought to have been formed by the removal during weathering of the softer Cedarberg shales and Pakhuis tillites, exposing the flat-lying upper beds of the more resistant Peninsula sandstones.

Following conformably on the Graafwater and Peninsula Formations, some 4 000 m of sandstones and shales of the upper TMG and the overlying, younger Bokkeveld and Witteberg Groups were deposited. With this pile of sedimentary material on top of them, accumulated over a period of tens of millions of years, it is not difficult to conceive of how the unconsolidated sediments of the TMG became steadily compacted and solidified into hard, resistant rocks. In many of the quartz arenites, the quartz grains became recrystallised and strongly cemented together to form the quartzites that are now exposed as the great amphitheatres and cliffs of the Western Cape mountain ranges. However, lithification (rock formation) took place as the TMG slowly sank deeper beneath the earth's surface, probably to depths of 4-5 km. It required geological processes of a completely different nature to expose these rocks once more as the high mountain ranges of the southern Cape.

THE MOUNTAIN-BUILDING EPISODE

About 250 million years ago, the cycle of continuous shoreline sedimentation was broken by a major event that disrupted the continental margin of the Gondwanaland supercontinent, buckling it into large fold structures and tearing and fracturing it along fault planes. One theory has it that this was caused by a small continental mass colliding with the southern margin of Gondwanaland in the South American-southern African-Antarctic region, in a similar manner to the collision of the Indian continental fragment with Asia some 150 million years later that resulted in the formation of the vast Himalayan mountain massif. In an area of continental collision, there is a tremendous amount of folding or buckling and movement of crustal blocks along nearly horizontal fault planes called thrust faults. At the same time, there is a great deal of uplift as parts of crust that were deeply buried are raised up to heights thousands of metres above sea-level, forming major mountain chains called orogenic belts.

In the case of the collision involving Gondwanaland, the major compressive force came from the south, buckling the flat sediments of the Table Mountain, Bokkeveld and Witteberg Groups into large-scale east-west folds. In certain anticlinal (convex-upwards) folds, rocks of the TMG rose to over 2 000 m above sea-level, where subsequent erosion of the softer overlying Bokkeveld and Witteberg sediments has exposed the TMG as the east-west mountain ranges of the Swartberg and Langeberg, the Riviersonderend and Outeniqua mountains, the Baviaanskloofberge and others. In the vicinity of the southern coast, the entire continental crust was uplifted and large areas of even the older basement rocks below the TMG were exposed above sea-level, particularly in the area between Cape Town, Saldanha Bay, Piketberg and Paarl.

Naturally, the creation of mountain-lands meant the termination of the Western Cape area as a basin of deposition receiving sedimentary material from the northern interior of Gondwanaland. Instead, it now (from about 250 million years ago) became an active area of erosion and weathering, with limited amounts of sediment accumulating in the inland Karoo basin to the north and more extensive drainage to a coastline to the south. Gradually, the overlying sediments were stripped and the sediments of the TMG once more exposed as mountain-lands well above sea-level.

This was not, however, the termination of major mountain-building activity along the southern margin of Gondwanaland, but rather the end of the compressional phase during which the folding and thrust faulting occurred. About 160 million years ago, the first major signs of the break-up of the Gondwana supercontinent appeared in the form of large-scale fractures called normal faults. In contrast to thrust faults where one block is pushed over the other by compressive forces, normal faults occur when one block drops or rises relative to the adjacent block, generally along vertical or near-vertical planes. The best known of the normal faults in the Western Cape is the Worcester Fault stretching from Tulbagh to Mossel Bay along the southern side of the Langeberg mountain range and having a measured vertical displacement of 7 000 m across the fault in the Robertson area. The effect of the faulting was once again to submerge large areas of the TMG beneath the sea along the margin of the newly-formed African continent.

On the Cape Peninsula, evidence of the normal faulting is relatively common on a small scale, and major faults are believed to cut across the Cape Flats and False Bay. From their measured orientations, the faults on the Peninsula fall into two sets or groups – a dominant northwest-southeast set and a minor northeast-southwest set. The Cape Peninsula can be subdivided into four fault-bounded blocks – a large northernmost block and a series of three smaller blocks making up the southern Peninsula. As one progresses southwards down the Peninsula, so, in general, are rocks of the TMG downfaulted closer to sea-level, ultimately becoming submerged beneath the sea on the western coast of the Cape of Good Hope Nature Reserve.

The first major normal fault approximately follows the Ou Kaapse Weg and separates a northern Table Mountain/Constantiaberg block from the adjacent Muizenberg Mountain block that is downthrown by about 100 m. The Muizenberg block is in turn separated from a southern Fish Hoek-Simon's Town block by the Fish Hoek-Noordhoek fault that lies along the length of the Fish Hoek Valley on the Clovelly side. The southern block is uplifted compared to the Muizenberg block, and exposures of Cape Granite reappear beneath Graafwater sediments on the southern side of the valley. This again indicates a displacement along the fault plane of greater than 100 m. The third major fault lies between Smitswinkel Bay and Die Mond in the Cape of Good Hope Nature Reserve. Here the Cape Point block has been downthrown by at least 150 m, relative to the Fish Hoek-Simon's Town block.

Several minor northwest-southeast faults cut across the back of Table Mountain itself, and many of the ravines and kloofs have formed by rapid erosion of the fractured sediments within these faults. Examples are Blinkwater Ravine, Kasteelspoort, Slangolie Ravine and Skeleton Gorge. Broken ground in the fault zones was commonly the site of percolation of groundwater that carried with it dissolved iron and manganese salts. These metals subsequently precipitated in cavities and along fracture planes as iron and manganese oxides. In certain places (such as the Hout Bay slopes of Constantiaberg and at Oppel's Kop on Devil's Peak) evidence of mining activities of manganese ores can be observed. Along certain faults, thin tabular bodies, called dykes, of a dark igneous rock, dolerite, have intruded into the Cape Granites and the TMG sediments. A fine example can be seen at Logie's Bay near Llandudno.

THE WEATHERING EPISODE

As the vast mass of Witteberg, Bokkeveld and upper TMG sediments was progressively eroded and stripped away from above the Peninsula and Graafwater Formations, the tremendous pressures that compacted the sediments into solid rocks were slowly dissipated. Rocks respond to pressure release by expanding and splitting along vertical and horizontal planes called joints. Joints are similar in appearance to faults, except that there is no movement between adjacent blocks on either side of the joint plane. Joints are far more prolific than faults in the sediments of the TMG, and the block-like outer physical appearance of the buttresses, pinnacles, amphitheatres and arches of the Peninsula mountain chain has been sculptured by joint-bounded slabs and blocks breaking away from the cliff faces during weathering.

Water also penetrates the Peninsula sandstones along joint planes, forming in places deep clefts, potholes, caves and subterranean streams. The Kalk Bay caves are fine examples of this selective weathering process. Water in the joint planes, like that in the large fault zones, also precipitated manganese and iron oxides on the joint walls. Fine manganese oxides crystallise in delicate dendritic growth patterns that have an intricate plantlike appearance and are frequently mistaken for plant fossils. When found in place and not on loose blocks, however, the dendrites can be seen to be crystallising on vertical joint planes that cut across the horizontal sedimentary beds.

The soils that are formed by the weathering of the Table Mountain sandstones are naturally dominated by fine quartz grains, and contain very few clay minerals and soluble salts that provide the basic nutrients for plant growth. In addition, organic acids break down the few clay minerals present, leaching important minerals out of the upper portions of the soil profile. As a result, the mountain soils are extremely nutrient-poor and acidic, and this has been one of the major controlling factors in the development of our rather subdued mountain fynbos flora. Unlike clay minerals, quartz grains have no adhesive properties and therefore erosion, once propagated, is extremely difficult to control in these sandy soils – particularly as regrowth of the vegetation is slow in the nutrient-poor environment. Peaty soils with high organic materials content have developed in marshy areas where the drainage is poor, such as the Back Table.

As the forces of nature bear down relentlessly on the mountain buttresses, so the Peninsula mountain chain is slowly and inevitably crumbling before the onslaught. Joint blocks break loose from the cliff faces and fall on the slopes below, collecting as angular boulders and rock fragments. In certain areas such as Slangolie Ravine and beneath Newlands Ravine, spectacular rock slides have resulted in large accumulations of boulders called screes. On the mountain slopes, the unconsolidated boulders and rock fragments roll and slump progressively lower and lower, grinding against one another, becoming rounder and smaller, and creating more and more fine sand and silt that are washed away in the fast-flowing streams. On reaching the sea, much of this sand has accumulated as the Cape's famed golden beaches. The ancient sedimentary rocks of the TMG have, in the process, been completely recycled and are now reforming as new sediments along our present coastline.

Climate

The climate in the Cape has varied dramatically in the geological past as evidenced by the appearance in the geological record of rocks such as the glacial deposits in the TMG. The modern-day climate, however, evolved during the Cenozoic geological era, from approximately 65 million years ago until the present. The evolution of the climate coincided with the opening of the Atlantic Ocean after the break-up of the Gondwana supercontinent and the gradual separation of the African and South American continents. At the time of the initial separation of the two continents, the Cape was situated further south than today, in the region of 50-55 °S. lat. After the break-up, the African continent moved steadily northwards, with the Cape reaching its present latitude of 34 °S. about 25 million years ago. At the same time, Antarctica split away from Africa and migrated to the polar regions, resulting in the formation of the circumpolar oceans and the growth of the continental polar ice-cap.

Approximately 40 million years ago, the Atlantic Ocean had opened sufficiently for the modern-day ocean current systems to start developing, and 10 million years ago, the Benguela current that brings cold, nutrient-rich ocean-bottom water from the Antarctic regions up along the western coast of southern Africa had come into existence. With the establishment of the cold Benguela current along our west coast came the stabilisation of the weather pattern around the southern portion of the African continent. This pattern includes a more or less static subtropical high-pressure cell over the South Atlantic and the regular passage of eastward-moving low-pressure cells to the south of the continent. The development of this stable weather pattern resulted in the creation of Mediterranean climates along the southern margin of the continent and associated arid desert regions further north. The arid regions along the west coast, including the Namib Desert, result from a combination of a virtually permanent high-pressure weather system and the upwelling of cold Antarctic water along the coast, creating moisture-free conditions in the upper atmosphere.

Since the stabilisation of the weather pattern in the southern Atlantic Ocean and adjacent continental areas some 10 million years ago, there have been large-scale fluctuations in the climate due to the variable nature of the earth's orbit relative to the sun. Over the past 2 million years, this has resulted in fairly extended periods of cool climatic conditions of about 100 000 years' duration, equivalent to the glacial periods in the northern hemisphere, separated by much shorter intervals of warmer conditions. The so-called glacial periods resulted in the growth of the Arctic ice-cap over much of northern Europe. There is no evidence that glaciers, as such, ever developed on Table Mountain or elsewhere in the southern Cape mountains at this time. The growth of the ice-caps in the Arctic regions, however, caused a general lowering of the level of the earth's oceans by tens of metres.

At the present day, we have experienced some 5 000 years of the most recent interglacial period and should at some stage in the relatively near future enter the next glacial episode. The last glacial period commenced about 120 000 years ago and was at its coldest approximately 20 000 years ago. During this time, the sea-level dropped to levels 130 m lower than at present, resulting in the retreat of the coastline to the mouth of False Bay. Large dune-fields covered False Bay and the Cape Flats, and from the orientation of these dunes it is obvious that the south-east winds had already become a dominant feature of the weather pattern in the Cape. It is estimated that the sea rose above the level of the Cape Flats during certain interglacial periods over the past 10 million years, and that, for brief periods, the Peninsula mountain chain existed as two separate islands, cut off from the mainland, with the sea covering the flats and also flowing through the Fish Hoek Valley. Raised marine platforms were created at this time by wave action, the best preserved examples being Green Point Common and an area surrounding Slangkop lighthouse on the southern Peninsula.

Nowadays, the Cape Peninsula experiences a Mediterranean-type climate with colder, wetter winters and warmer, drier summers, neither of which are too severe. The climate is essentially controlled by two major atmospheric circulation features of the southern hemisphere – the South Atlantic anti-cyclonic high-pressure zone and a circumpolar belt of eastward-moving, cyclonic low-pressure cells to the south. In winter, the South Atlantic anticyclone moves some 5 ° towards the equator, allowing the westerly low-pressure cells to brush the southern continent, bringing moisture and cloud associated with west to north-west winds.

While these are the principal constraints on climatic variation in the Cape, there are several more local controls that have a significant effect on the local weather patterns. The most obvious of these is topography, or more specifically, the presence, height and shape of various mountain masses. Another factor is the presence of warm Indian Ocean water in False Bay and the cold Benguela current off the Atlantic shoreline. Combinations of local factors can result in considerable variations in climatic elements such as temperature, rainfall, wind and cloud cover over the Peninsula.

The mean annual temperature on the Cape Peninsula is approximately 17 °C. This mean does not vary greatly from month to month, and, due to the moderating effects of stable adjacent sea temperatures, the daily range in temperature is generally not excessive. With increasing altitude in the Peninsula mountains, there is an approximate decrease in temperature of 0,5 °C per 100 m of altitude gained. The average maximum midsummer temperatures on the summit of Table Mountain are just less than 30 °C, and the average minimum midwinter temperatures just over 2 °C. At the Kirstenbosch Gardens, these average temperatures are some 3 ° higher. Maximum temperatures are normally attained in the early afternoon (14:00) and minima in the early morning (06:00).

The maximum possible hours of sunlight vary from 14,5 hours in midsummer (05:30 to 20:00) to a minimum of 10 hours in midwinter (07:45 to 17:45). The hours of sunlight experienced on the mountain, however, can be reduced due to frequent cloud cover. On average, 184 days of mist (one day in two) per year are experienced at the summit of Table Mountain. The attitude or angle of the mountain slopes also has an effect on temperature, with the sunny northern and western slopes experiencing hotter conditions than the eastern and southern slopes. This phenomenon is enhanced in winter when the sun is at a lower angle. Walks on the front face of Table Mountain can be exceedingly hot in summer but pleasant in winter, and likewise routes up the eastern slopes from Nursery to Newlands Ravine can be cool in summer but cold and wet in winter.

Rainfall is another element of the climate that is controlled to a large extent by the topography and elevation of the mountains. Rainfall over the Peninsula generally occurs because of a combination of two factors. The first is cyclonic activity associated with the low-pressure cold fronts in the circumpolar belt of westerlies. Within the low-pressure zones, mixing of converging masses of cold air from behind the front with warmer air preceding the cold front causes condensation and the formation of moisture-laden cloud belts that accompany the eastward-moving cells. The second feature is the enhancement of condensation and rainfall by further cooling of the moisture-laden air as it is forced to rise over the mountain ranges. As a result, most of the rainfall experienced over the Peninsula is precipitated on the mountain chain itself, or in close proximity to the mountains.

The annual recorded rainfall on the summit of Table Mountain is in excess of 1 900 mm, and that on most of the high-lying mountains down to Constantiaberg and the Muizenberg mountains in excess of 1 000 mm. In contrast, the lower-altitude areas of the southern Peninsula from Fish Hoek to Cape Point have annual rainfall figures of about 700 mm, and the Cape Flats, lying in the rain shadow of the Peninsula mountain chain, only 400 mm. About 80 per cent of all the rain falls in winter, with precipitation recorded on between 120 and 140 days per year in the summit region. At least twenty cold fronts sweep across the Peninsula each year, with the heaviest falls generally occurring at night or during the early morning. Sudden drops in temperature in the region of 5 °C are experienced with the passage of cold fronts, that commonly last from 2-4 days. Light snowfalls are occasionally recorded in the summit area, averaging about one fall per year with the snow seldom surviving for longer than 24 hours before melting.

Due to the regular presence of enveloping mists on the mountains, the amount of moisture precipitated is in fact considerably greater than the actual recorded rainfall. While 1 900 mm falls at the mountain summit, an additional 1 300 mm precipitation occurs annually by means of moisture absorbed directly by plants as they make physical contact with the swirling mists.

The Cape Peninsula is characterised by strong winds that are associated with the major weather patterns. In summer, the anticyclonic south Atlantic high-pressure system frequently extends to the south of the Cape, creating south and south-east winds over the Peninsula. The general north-south alignment of mountains on either side of False Bay has a wind-tunnel effect and increases the velocity of these winds across the bay and the Cape Flats, particularly in close proximity to the mountains themselves. The Table Mountain massif, on the other hand, diverts the south-east winds and creates a relatively wind-free zone in the lee of the mountain that includes Lion's Head and Signal Hill and extends out across Table Bay.

South-east winds dominate the weather pattern for about 50 per cent of the year, and north-west to west winds, associated with the cold fronts, for about 30 per cent of the year. On the Peninsula mountain chain, however, the irregular nature of the topography can have radical effects on the direction and speed of the winds. Velocities increase dramatically as winds pass over mountain ridges or crests such as the top of Table Mountain, and complex gusts and eddies are created on windward and leeward slopes. The spectacular, swirling tablecloth over Table Mountain is formed when strong south-east winds, bringing cool, mist-laden air in large, swirling eddies over the top of the mountain, meet warmer air rising within the city bowl, and rapid evaporation of the leading edge of the cloud mass occurs.

For about 20 per cent of the year, most commonly in the late summer and autumn when the cyclonic winds are least prevalent, a sea and land breeze circulation is established due to the more rapid heating and cooling of the land mass during the day and night respectively, compared to the relatively constant temperatures over the sea. During the day, hot air rises over the Peninsula, causing a sea breeze over the southern Peninsula, and, simultaneously, north-west winds over Table Bay.

At times, the sea breezes can bring with them sea fogs from the Atlantic Ocean. These fogs form when air overlying warmer water further offshore moves landwards over the cold upwelling inshore Benguela current and condenses. On average, about 75 days a year have associated sea fogs in the Cape Point area, although the number of days when these fogs invade the northern Peninsula is considerably less. The fogs are generally nocturnal to early morning phenomena, and rapidly disperse before midday. On still nights, land breezes may occur as the mountains cool rapidly. These breezes are generally very light and are mainly experienced in the larger ravines or valleys.

The weather pattern in the interior of South Africa is dominated by a second anticyclonic high pressure cell that is commonly situated over the central Karoo. Hot, dry berg winds associated with this cell occasionally intrude on the climate at the Cape from the north or north-east. Strong berg winds can account for dramatic temperature increases (up to 10 °C) and, together with the summer southeasters, represent periods of particular fire hazard in the fynbos vegetation. Rainfall in the interior occurs together with thunder activity, and in the summer impressive cumulus clouds can be observed from the Peninsula mountains, gathering further inland. It is only on exceptional occasions, however, that these thunderstorms are experienced over the Cape Peninsula.

All these climatic variables interrelate to make the Cape Peninsula mountain chain an area of markedly differing environments and ecological niches, and this, in turn, has led to the establishment of a wide diversity of flora and invertebrate fauna. The climatic variability is also reflected in constant and sometimes rapid changes in the weather, and the hiker venturing onto the mountains should be well equipped for most weather conditions.

Acknowledgements

Prof C J Hartnady, Precambrian Research Unit, University of Cape Town.
Dr J Rogers, Marine Geoscience Unit, University of Cape Town.
Mr B Gasson, School of Architecture and Planning, University of Cape Town.

Flora of the Cape Peninsula mountains

PROF W P U JACKSON

"Oh, Lord, how many, how rare and how wonderful were the plants that presented themselves to Hermann's eyes," the great Swedish naturalist and taxonomist Carolus Linnaeus wrote in 1747 after examining the collection made some 75 years earlier by the early Dutch explorer Paul Hermann, the first professional botanist to gather plants at the Cape. In 1822 the English traveller and naturalist William Burchell, after a walk on Lion's Head, wrote: "All that I had pictured to myself of the riches of the Cape in botany was far surpassed. At every step a different plant appeared; and it is not an exaggerated description if it should be compared to a botanical garden, neglected and left to grow to a state of Nature; so great was the variety everywhere." Such reactions are not unusual, and have been echoed by many others over the years, whether scientists or just ordinary men and women. The Cape is, in fact, the home of the smallest but richest of the six floristic kingdoms on earth. Commonly known as fynbos, the flora of the Cape Peninsula includes some 2 600 species.

What is fynbos?

As its name suggests, fynbos is largely composed of low, shrubby vegetation with small, tough, narrow leaves, almost entirely evergreen, adapted to withstand periodic drought and often waterlogging as well. Trees are few in mountain fynbos. The most characteristic families are PROTEACEAE, ERICACEAE and RESTIONACEAE (reeds). Fynbos is the most important constituent of the Cape Floristic Kingdom. The Cape Peninsula is a typical sector of this kingdom, being unique in a number of characteristics.

The Cape Floristic Kingdom is extremely rich in numbers of plant species – in fact, the Western Cape is possibly the most species-rich area in the world. It contains enormous variety, with a different species found with every step, and an almost complete lack of single species dominance. Even in an apparently homogeneous stand of pink ericas you will find many other species – restios, agathosma, asparagus and so on.

There are a large number of endemic taxa, that is, families, genera and species not found anywhere else – perhaps the highest endemicity in the world. The Cape Peninsula does not contain any genera unique to itself but it contains several endemic species, including *Mimetes fimbriifolius*; *Serruria vallaris* (*S. villosa*), a low, bushy, proteaceous species with yellow flower-heads, common in the Kalk Bay mountains; two other serrurias, one being *S. collina*, rediscovered recently by members of the Botanical Society after being lost for 200 years; *Aloe commixta*, a yellow-flowering, rock-hugging aloe from the southern parts; two striking staavias (diamond eyes) of the BRUNIA family, with leafy, starry white heads; several ericas, including *Erica sociorum*, *E.*

Fynbos on the plateau above Kalk Bay. (*Photo: W P U Jackson*)

urna-viridis, *E. fairii* and *E. gilva*; *Gladiolus aureus*, the golden gladiolus, a highly endangered species, existing in a single small colony; *G. vigilans*, keeping its lonely vigil at Cape Point; and finally, Table Mountain's own gladiolus, *G. monticola* (meaning "arising in the mountain") – to mention just some of the more showy species.

Within several genera an "explosion of species" has occurred, producing more species to a genus than are found in other floras. We have mentioned the largest genus, *Erica*, with about 100 species; others are *Senecio* (the daisy family) with 44 species; *Aspalathus* (the pea family) with 42; *Ficinia* (the sedge family) with 33; *Oxalis* (the sorrel family) with 33; *Crassula*, with 30; with several others close behind. These figures refer to the Cape Peninsula only.

Frequently, relatively rare species occur in disjunct populations, widely separated from each other. A good example within our area is the extraordinary marsh-loving *Witsenia maura* (bokmakierie's tail) of the IRIS family. Several striking species occur in the southern part of the Peninsula and then again around False Bay, 70 km or more away. These include the beautiful *Mimetes hirtus* (red and yellow bottlebrush) found in the

Bokmakierie's tail (*Witsenia maura*). (*Photo: W P U Jackson*)

Cape Point Nature Reserve and at Betty's Bay; another proteaceous species, *Diastella divaricata*, a low straggly bush carrying small pink pincushions; *Erica patersonia*, the popular mealie heath; *Herschelia purpurascens*, the early purple disa; *Gnidia penicillata*, with small, starry blue flowers; and *Audouinia capitata*, the false heath.

The latter is a rare and striking plant with bright carmine flowers, best seen on Karbonkelberg but occurring also in six other tiny enclaves of one to sixty plants only. It is the only species of the genus in the fynbos-endemic family BRUNIACEAE. Almost endemic to the Peninsula, another colony has recently been rediscovered 90 km to the east! The authoritative *Red Data Book*, which reports the find, declines to state where it is. Unless there is evidence for dispersal of seed over long distances, e.g. by birds, then the disjunct distribution of rare species suggests that they are ancient, previously possessing a much wider range, and may therefore be regarded as relicts. Apparently *Audouinia* does not set seed at all.

Altitude and aspect cause more frequent changes of plant communities in mountain fynbos than occur on other mountains. Thus the flora of the Lower Table of Table Mountain (about 800 m) differs enormously from that of the lower slopes (about 400 m) and of the Upper Table (about 1 000 m). The same applies *a fortiori* to the difference between the northern and southern slopes.

Trees are scant in mountain fynbos. The protea family supplies some, including *Protea nitida* ("waboom"), *Leucospermum conocarpodendron* ("kreupelhout"), *Mimetes fimbriifolius*, and, most particularly, the silver-tree, *Leucadendron argenteum*. *Widdringtonia nodiflora*, the mountain cypress, is our only cedar. It is easy to recognise, with its tiny triangular leaves and its long-held cones. If unburnt, it grows into a typical cedar shape. A few other small trees may be found in damper patches on the slopes, including *Maytenus oleoides* ("klipkershout"), which resprouts from its upper branches after being burnt (just like *Protea nitida*); *Podalyria calyptrata* ("keurtjie"); *Maurocenia frangularia* (Hottentot's cherry); *Phylica buxifolia* (box hard-leaf), which has strongly scented flowers in autumn; *Erica caffra* ("waterheide"); some *Rhus* spp ("taaibos"), with their three leaflets; and *Colpoon compressum* (Cape sumach), with its edible purple berries.

The monocotyledons – i.e. bulb-type plants and grasslike families – are unusually well represented in our fynbos, though grasses are relatively few.

One of the most remarkable features of fynbos is its relation to fire. Not only has fynbos adapted to fire; it has evolved with fire and from fire. Much of it encourages fire, and it could not exist without fire. Fire does

not destroy fynbos; it rejuvenates it. But the fires it grew up with were infrequent (caused by lightning, earthquakes and rockfalls), uncontrolled, vast in extent and at variable intervals. Consider our fire-lily *(Cyrtanthus ventricosus)*, a pink amaryllid that flowers within 4 weeks of a fire but never without one; or our Barberton daisy (*Gerbera asplenifolia*), with attractive, fern-like leaves, whose white or pink flowers occur only after a fire. Imagine the evolutionary process that brought about such a phenomenon, and that without fire such plants must have lain dormant for up to 100 years or more.

Many other species flower only, or far more prolifically, after a fire, especially those in the bulbous and cormous lily, iris, amaryllis, bloodroot and orchid families, and the daisy and pea families. Restios produce exuberant new growth. The tall pink *Watsonia borbonica* (*W. pyramidata*, "suurkanol") burgeons into massed colour perhaps 10-12 months after a fire.

The fynbos thus differs radically from other, non-fire-dependent biomes by being in constant change; it differs from European plant communities in that it is limited by time as well as space. In the first year after a fire we find grasses and geophytes (plants with dormant undergound parts only, such as bulbs), and a number of daisies such as gerberas, alciope and haplocarpha ("brandbossie") – the last two being what I call post-fire Table Mountain daisies, flowering respectively in summer and winter. Several othonnas, also with yellow flowers and shiny, smooth leaves, bloom from protected underground tubers. The attractive, mainly pink-flowering corymbium ("heuningbossie") is also a daisy but has grasslike leaves. Some restios produce exuberant growth, but will only flower later.

A few ericas, particularly *Erica cerinthoides*, flower early from fire-resistant rootstocks. Most ericas, regrowing from seed, need several years to flower and may dominate the vegetation for some time, but eventually they will be replaced by the so-called "climax vegetation", characterised by the protea family but including shrubs like *Passerina* spp ("gonnabos"); the ericoid *Philippia* ("kabinethout"); cliffortias; myricas (waxberry); *Athanasia* spp ("Klaaslouwbossie"); some of the larger pelargoniums; straggly asparagus and "wag-'n-bietjie". Peas (family FABACEAE) appear quite quickly after a fire. The genus *Aspalathus* is particularly well represented; after a few years they will become "leggy", then senescent, and die off.

So what you see in flower one year will probably not flower the next year. The climax vegetation in fynbos seldom becomes senile (about 30-40 years old) because it is usually burnt again before that time. Certainly there are some areas where this sequence, the "post-fire succession", gets held up, as on the damp flats in the Cape Point reserve which are suitable for restios but not for the later successional species. Thus, what you see when you walk through fynbos is the product of the species alive in the soil times the duration since the last fire.

Of course fire can be destructive – when it is too hot, too frequent or occurs at the wrong time of year. Many man-made fires are all of these. We do not really know the ideal frequency of fires; obviously the slowest maturing plant must have time to produce a large enough crop of seed to ensure its survival – 15-20 years for some members of the protea family. An autumn burn, at least for proteas, seems optimal: their seed needs the winter rain for sprouting; if it lies about in summer, it gets eaten or loses its viability. But this may be wrong for other

Haemanthus in bloom after a fire. (*Photo: W P U Jackson*)

The "suurkanol" (*Watsonia borbonica* = *W. pyramidata*) above Hout Bay. (*Photo: W P U Jackson*)

A protea field after a fire. (*Photo: W P U Jackson*)

plants – if we want to preserve them, especially certain geophytes, should the burning be more frequent? It does seem reasonably certain that burning (e.g. of fire-breaks) in cycles of less than 10 years is potentially destructive. Perhaps the cycles should be irregular; perhaps the fynbos would be better off without fire-breaks at all.

How did fynbos develop?

What has given rise to the development of this unique flora? The reasons are complex and not fully understood. The South African botanist Hugh Taylor reduces these factors to one word. It is the vicissitudes, he says, that have encouraged speciation, radiation and hybridisation in the Cape – major among them being the climatic fluctuations, the changes in sea level, seasonal droughts and floods, and nutrient-poor soils.

The entire fynbos biome is virtually an island. Ancient temperate and tropical floral elements have been able to undergo divergent development because it has become increasingly isolated since the Palaeocene – cut off by the sea to the south and west, and by a belt of increasingly arid land to the north and east. The Cape Peninsula is obviously a sort of "island within an island", almost totally isolated by sea and sandy flats.

An important reason why our flora is so much richer than those of temperate Europe and North America lies in the almost total destruction of the northern flowering plants by the successive ice ages of the Pleistocene, which only finally retreated some 10 000 years ago. The last ice age that directly affected southern Africa was over 300 million years ago, following the deposition of the final layer of Table Mountain sandstone. Since flowering plants did not develop for another 200 million years we can see that our vegetation has never suffered a comparable devastation. There is no doubt that our flora is a very old one (though recently expanding), while that of the northern areas is very young.

The home of fynbos is Table Mountain sandstone – the soil it produces is just about the most nutrient-poor in the world. It is particularly low in nitrogen and phosphorus, acidic, generally shallow, highly leached and porous (easily drying). This very inhospitability encourages adaptations and hence speciation.

The Cape region possesses an extremely dissected landscape with numerous mountain ranges and drier intervening valleys and flats. This produces a whole array of restricted and varying niches, suitable for highly variable plant species.

The Mediterranean-type climate (with winter rainfall) that prevails in the south-western Cape probably arose only in the Pleistocene and encouraged speciation in an already complex area. The repeated fluctuations of temperature and moisture during the Pleistocene, together with the topography mentioned above, led to recurrent isolation of plant populations, and stimulated geographical speciation.

The major components of fynbos

THE PROTEA FAMILY

"What on earth is that?" my wife and I exclaimed almost simultaneously when we first sighted the giant protea (*Protea cynaroides*) on Table Mountain shortly after our arrival in South Africa in 1949. How could such a large, striking, symmetrical object be a wild flower? What a contrast to the small British field and hedgerow daisies, roses and violets! In fact, in addition to its size and grandeur our protea differs in many ways from the typical northern hemisphere plants and may be considered to exemplify the vegetation of the fynbos of the Cape Floristic Kingdom. To start with, that pink head is not a flower but an inflorescence, a collection of hundreds of florets, each one with its own four petals, its unobtrusive stamens and its peculiar style, known as a pollen presenter, which stands erect in the centre of each opened flower. The large pink petals are not petals but floral leaves, or bracts, which surround all the florets. Inside the heads many insects may be seen, attracted by the nectar and the pollen, and the larger beetles are probably important in effecting pollination. The flowering season is primarily late summer and autumn on Table Mountain, but in other areas flowers occur at different times of the year.

The leaves are also unusual, being broad, firm and tough, and resist predation by containing unpleasant phenols and terpenes. The upper and lower surfaces of the leaf are alike (isobilateral) and therefore photosynthesis can occur on either side; in fact, in the protea family as a whole the leaves tend to stand upright rather than flat, so presenting their edges to the hot sun. Their roots are also odd in that they carry immensely fine and numerous rootlets, presumably allowing them to absorb as much food as possible from the extremely nutrient-poor soil. The plants are also resistant to fire – after the aerial parts have been completely destroyed, fresh sprouting occurs within a few months from the thick, tough "ligno-tubers" which grow at ground level. So a plant under a metre tall may be many, perhaps hundreds, of years old, and may have been burnt many, many times.

The first protea ever to be illustrated, in 1605, was believed to be a thistle, while *P. cynaroides* was likened to a globe artichoke. Linnaeus then chose the name Protea for the genus after the god Proteus, who could adopt many different forms, and the specific name was constructed from cynar, artichoke, and -oides, like. It is aptly called the giant or king protea, and it is our national flower.

The idea of the protea resembling a thistle is legitimised in the scientific name of the charming little yellowish-green *P. scolymocephala* – or thistle-head – which grows in the central part of the Peninsula. Then there is one ground protea, *P. acaulos* (Greek for "without stem"), with leaves and flowers really sitting on the ground, common throughout the area but not so often seen in flower. This protea is almost certainly pollinated by small rodents, which visit it at night, attracted by its yeasty odour.

At the other end of the scale is the tree protea, *P. nitida* (meaning "shining"), or "waboom" – so called because its softish wood was used for making brake blocks for the old Cape wagons. Colonists made ink from the leaves and a rusty nail. This protea is one of the few trees – all rather small – found in fynbos on open slopes, and shows a further adaptation to fire in that it can resprout not only from the base but also from aerial branches which carry deep-seated, well-insulated, "epicormic buds" that are stimulated to shoot after a fire. *Maytenus oleoides* ("klipkershout") is another small tree with the same property, also shared by the Australian gums, as can be clearly seen in the plantation immediately south of the botanic gardens at Kirstenbosch.

Tree protea (*Protea nitida*). (*Photo: D Hey*)

Protea lepidocarpodendron, the black-bearded protea, does not resprout after being burnt but regrows from a massive seed reserve, sometimes forming dense forests in the Chapman's Peak area. These look at first glance like a monoculture, a single species dominance that is uncommon in the fynbos, but this is in fact more apparent than real.

We cannot leave the proteas without mention of *P. repens*, our beloved sugar-bush, or "suikerbossie". This pink and white protea once clothed much of Lion's Head and the eastern slopes of Table Mountain, but has now been exterminated from these and from most flat areas by fires, picking and "development". It still exists in the Llandudno area and in central parts of the mountain chain, up to the 350 m contour occasionally in dense communities. This bush was aptly named *P. mellifera* (meaning "honey-bearer") by Thunberg and this remained its name until recently. But apparently Linnaeus was originally misled by Boerhaave's sketches into believing it was of creeping habit, and in 1753 used the epithet repens. So the name has had to revert to the original, according to the laws of nomenclature. The sweet, sticky nectar from the perianth of this protea, named "bossiestroop", has been popular as a remedy for coughs, chest diseases and even diabetes.

Proteas have developed a further fire-related strategy, called serotiny, in which they keep most of their seed in persistent heads that maintain viability for several years, only being released after a fire. Many leucadendrons do the same, as do some other unrelated fynbos genera, particularly the brunias and the invasive Australian proteaceous hakeas.

Other genera of the protea family on the Table Mountain range include *Leucospermum* (pincushions), *Leucadendron* ("geelbossies"), *Mimetes* (meaning "mimic"), *Diastella*, *Serruria*, and *Brabejum* (wild almond). *Leucospermum conocarpodendron* ("kreupelhout") is another small tree found on lower open slopes. It is a common sight all over the Table Mountain range, but is particularly prominent from the Pipe Track above Camps Bay. The furry leaves are tipped with several bright red, firm and swollen teeth which are believed to be "extra-floral nectaries" and are characteristic of many proteaceous genera.

"Pollen presenters" constitute another unusual feature of the hermaphroditic protea genera, and stand out particularly well in the pincushions – they are in fact the pins. In each individual flower the simple style elongates more rapidly than the petals and becomes held at the tips like a taut bowstring. The flower snaps open, the style bursts free of the petals and in so doing brushes off pollen from the anthers which actually sit on the petals near their ends. The style, having received the sticky pollen, becomes a "pollen-presenter". The petals now generally collapse, leaving the pollen-presenters "in charge". The pollinator, whatever creature it may be, must push past these pollen-presenters to get at the basal nectar. It then carries the pollen grains to the next inflorescence which may be a little older, so that a receptive stigma can receive the pollen.

Leucadendrons differ from other proteaceous genera in being dioecious, i.e. carrying male and female flowers on separate bushes and having its seed in a cone. Like the sugarbush, the silver-tree (*Leucadendron argenteum*) was previously far more abundant on Lion's Head and Devil's Peak, and, shimmering in the breeze, must have presented a fabulous sight. It is now considered to be vulnerable in the wild, being confined to a granitic soil which occurs only at low altitudes and thus perilously close to several habitat pressures. It is seen on the lower western slopes of Lion's Head and at Kirstenbosch, where it should be safe – except perhaps from the pathogenic fungus *Phytophthora cinnamoni*, against which we do not yet have adequate protective measures. *Leucadendron xanthoconus* (meaning "yellow cone") is a real "geelbos", the upper leaves turning bright yellow at flowering time in winter and painting the upper mountain slopes with great splashes of colour.

The female cone of the silver-tree (*Leucadendron argenteum*). (*Photo: W P U Jackson*)

Mimetes fimbriifolius (meaning "leaf with silky hairs on the margins") is another small, proteaceous tree of the open slopes, endemic to the Table Mountain chain. When not in flower it is easily mistaken for the kreupelhout. When in flower the floral leaves turn a pinkish red in contrast with the yellow pincushions of the leucospermums, and the true flowers are largely hidden.

The Cape almond (*Brabejum stellatifolium*) is a true forest tree, more closely related to some Australian members of the family, and was used by Van Riebeeck to form the boundary of his settlement in 1660, partly to prevent the Hottentots from stealing cattle. A remnant of Van Riebeeck's hedge can still be seen at Kirstenbosch and on Wynberg Hill and is now a national monument. The Cape almond is unlike other Cape members of the family and not easy to recognise as one. The leaves

"Geelbos" (*Leucadendron macowanii*). (*Photo: W P U Jackson*)

Silver mimetes (*Mimetes argentea*). (*Photo: W P U Jackson*)

are toothed, lance-shaped and in whorls of six, while the fruit is an ovoid drupe covered with velvety brown hairs, resembling an almond. It is poisonous, containing hydrocyanic (prussic) acid. However, this can be removed by soaking and the nut can then be used as a coffee substitute. "Brabejum" is the Greek name for a prize awarded at the Pythian games – a crown of laurel leaves, somewhat resembling the leaves of the Cape almond.

The protea family is an old one. It is believed to have originated some 100 million years ago in the tropical region of the Australian portion of Gondwanaland, the great southern continent, which was in the process of breaking up, having already cast off India and Antarctica. So Australia, Africa and South America drifted apart, all carrying with them early proteaceous genera.

While South Africa is renowned for its variety of protea, Australia exceeds us by far in the number of genera belonging to this family, including forms differing widely in appearance such as the banksias, hakeas, dryandras, smoke bushes (*Conospermum*), grevilleas, isopogon and the sand plain pear (*Xylomelum*), some 750 species in all as compared to our 500 species. The rich variety that occurred in the course of time was partly related to fire. Some authorities go so far as to say that several species actually encourage fires by building up poorly decomposable, highly inflammable litter. After a fire the proteaceous species survive while many other groups are eliminated, and they benefit from nutrient release, death of parasites and re-establishment of short-lived species.

The protea family illustrates nature's favourite shape, the spiral. Look particularly at the colourful surrounding bracts in the bud and at the receptacle of the dead flower-head of any protea. In each case you see two sets of equiangular spirals, one rotating clockwise, the other anticlockwise. Each member, bract or individual seed-bed forms part of one clockwise and one anticlockwise spiral. The number of clockwise to anticlockwise spirals is always in the ratio 0,618, or as near to it as possible (e.g. 21:34 in the waboom seed-case).

The long-tailed Cape sugarbird (*Promerops cafer*) is closely associated with proteas and does not seem to be able to exist without them. It is a pollinator of proteas and leucospermums. The three local sunbirds are pollinators of leucospermums and mimetes, while leucadendrons are largely wind-pollinated. The family is designed for outcrossing, the stamens maturing before the stigmas, and many species are self-incompatible.

THE ERICAS

We have devoted much time to the proteas because they are so prominent and typify many features of the fynbos. Another characteristic family is that of the ericas (heaths), ERICACEAE. In the Cape this family consists of over 600 species of the genus *Erica*, together with several smaller and comparatively insignificant genera. Ericas can be defined as woody plants, ranging from a few centimetres to several metres in height, sepals and petals in fours, eight free stamens, anthers two-celled, opening by pores (characteristic), ovary superior, at least four-chambered, with two or more ovules in each chamber and fruit a small capsule.

The name of this genus is derived from the Greek word *ereiko* meaning "to break", used by Theophrastus, either because of the medicinal use of ericas to break up bladder stones, or, more prosaically, because the stems are brittle and break easily. The leaves in most species are tough and narrow, each leaf folded to produce a furrow that runs along its underside. (Such leaves are found in many other unrelated families in the fynbos and are known as ericoid – an illustration of the convergence of characteristics during evolution brought about by common environmental pressures.) This type of leaf reduces water loss; in fact, among the few ericas that inhabit damp and shady places the leaves are often flat or "open-backed".

About a hundred species of erica grow in our area; more species than in any other genus. Only a dozen species are found outside Africa. The origin of this family certainly differs from that of the PROTEACEAE, and is in fact uncertain. What is certain is that once the genus *Erica* had arrived in this area, its rapidly occurring genetic variation threw off new species at a rate probably unparalleled in any other family anywhere else in the world. This ericoid explosion has been fairly recent, however – within the last million years; perhaps the ge-

Erica coccinea. (*Photo: W P U Jackson*)

Prof Jackson's heath (*Erica jacksoniana*). (*Photo: Dolf Schumann*)

nus had been around long before, waiting for just the right environmental conditions.

It is not surprising then that some species of *Erica* are rare, even extremely so – as for instance *E. sociorum*, a greenish flower probably confined to a single rock-band on Noordhoek Peak. Another green flowerer is *E. urnaviridis* (meaning "green urn"), which is confined to the Kalk Bay range – but although local, you will find this one in its thousands alongside paths in the area. Do look out for it; it is fairly tall and leggy, with sticky, green, urn-shaped flowers. Comparable is *E. fairii*, known only from three or four small colonies some short distance from each other, but flowering nicely in the botanic gardens at Kirstenbosch.

On the other hand, some ericas are quite widely dispersed. *E. cerinthoides*, for instance, extends into the Northern Transvaal. This very popular heath bears fat, tubular, hairy, brilliant vermilion flowers throughout the year. It is highly resistant to fire – in fact, it is stimulated by fire to flower from its old rootstocks, and is one of the first flowers to be seen after a burn.

Comparatively few ericas resprout after a fire – most have to regrow from seed – but *E. mammosa* (ninepin heath) is one that does. Mammosa is derived from "mamma", or breast, but the comparison seems far-fetched, though it is also known as nipple heath. This is an erect, robust shrub with showy flowers. The curved, tubular corolla occurs in dull red, scarlet, pink, mauve and orange-red. A pale green variety sometimes known as *E. gilva* ("groenheide") may appear in dense communities, especially in the Constantiaberg area. Four indentations can be found at the base of the corolla and are helpful in identification. Flowering occurs mostly in midsummer.

Two other common, erect, robust ericas with tubular corollas are *E. plukenetii* (named after an English botanist, Plukenet), "hangertjie", and *E. coccinea*, meaning "red", though this is not its most common colour. Both have prominent protruding (or "exterted") brown stamens which tend to give them a rather untidy appearance. *E. coccinea* is distinguished by its clasping sepals and bracts which enclose the base of the corolla and are believed to prevent insects from stealing nectar.

Another tough, bushy heath is *E. abietina*. The flowers are sticky and the corolla is tubular, curved and red. It is confined to Table Mountain and Devil's Peak and flowers profusely in midwinter, when there is little else. It enlivens the dry, rocky ledges on the northern side.

These last five ericas are all sturdy plants with colourful, tubular corollas and are suitable for pollination by birds, especially the orange-breasted sunbird, *Nectarina violacea*. The other ericas mentioned above are probably all insect-pollinated. Those ericas with small and insignificant flowers are often wind-pollinated; such plants produce clouds of pollen and generally possess a comparatively large stigma, the better to collect it. If you walk into such a pollen-cloud, it is probably from *Erica hispidula* (meaning "little hairy one").

E. hirtiflora (meaning "hairy flower"), which is very common on the damp slopes, carries numerous flowers with pink, hairy, egg-shaped corollas. It is a social heath, as anyone can see from the splashes of pink that occur twice a year above and around the gardens at Kirstenbosch. It has been slowly spreading across the face of the mountain from Constantia Nek, moving further after each fire. On the other hand the rather similar and popular *E. mauritanica* ("bosluisblommetjie") seems to be receding. This heath is not hairy and can be readily identified by the rounded interstices between corolla lobes or "petals".

Another common, small-flowered erica is *E. lutea*. The name means "pale yellow" and that is its colour. It is a "cup and saucer" heath, with a cup- or urn-shaped corolla sitting in a prominent, saucer-like calyx. Its common name is "geelrysheide".

A few ericas grow into small trees; most prominent here is *E. caffra*. This dull, white-flowering, scented erica grows only near water – hence its common name of "waterheide" – and it is closely related to the Mediterranean tree heaths, from the boles of which the best briar pipes are made. Some ericas can present massed displays of pink and purple. This is probably *E. pulchella*, with deep pink, urn-shaped flowers carried in terminal spikes in late summer. Two other species are common, especially in the southern parts. One is a smallish, pink "cup and saucer" erica with a long flowering season, *E. corifolia* (meaning "leathery leaves"). It is characterised by the corolla lobes soon closing and turning brown. The other has a rather strange appearance, the almost globular flowers being densely clothed in long, fine, silky, spreading, pink and white hairs. This is *E. bruniades* ("kapokkie"), flowering from July to January.

Prof Jackson viewing Cape reeds. (*Photo: Dolf Schumann*)

THE CAPE REEDS

The third family that characterises the fynbos is the RESTIONACEAE or Cape reeds, familiarly known as restios. The name of the family is derived from the Latin word *restio*, meaning rope-maker, but the most important uses are for thatch and brooms. The fresh green shoots after a fire (restios are resprouters) make good fodder. To quote Pillans, they are "the most characteristic plants of the Peninsula landscape". Some species occur in great sheets, covering mountain and plain with shades of greenish-brown, rippling in the wind like a field of wheat. They are one of the grasslike (graminoid) families and indeed largely take the place of the grasses we would expect to find on mountains elsewhere. Their distribution is interesting in that they are confined almost entirely to South Africa and Australia.

The best places to see them *en masse* are at the extremes of our range – on the Upper Table of Table Mountain, and on the flats in the Cape Point Nature Reserve. Both these are damp and apt to be waterlogged in winter; many restios being particularly adapted to withstand this hazard and so grow where other plants cannot. These reeds were of inestimable value to the early settlers, who had no other roofing materials, and are still important for thatching today. One of the best thatching reeds is actually called *Chondropetalum tectorum*, derived from the Latin *tectum*, a roof.

All species in this family are dioecious, that is, the male and female flowers occur on different plants and the two sexes usually do not resemble each other, so that matching them may be difficult, although the vegetative parts are alike. Members of the family are perennial, usually tufted herbs, with the leaves reduced to small dry sheaths so that photosynthesis takes place in the green stems, which in some species are much branched and mostly sterile. The small flowers are borne in "spikelets", each one with six dry, brownish perianth segments ("petals") and three prominent stamens in the male flower. Wind-pollination is the rule, but small insects may play a minor role.

Restios provide an extremely good example of seed dispersal by ants, which is known as myrmecochory. Many of the genera provide their seed with colourful rounded succulent appendages, or eliasomes, that appeal so greatly to ants that they carry them off to their hiding-places, eat the titbit and leave the seed buried in the ground. So the seed remains intact, out of harm's

White everlastings (*Helichrysum vestitum*), Cape Point.

way beneath the surface, in the right place for sprouting and away from its parent. Ant dispersal occurs also in the protea family – it was first observed in the fynbos in *Mimetes*, and is probably important in many other families.

Individual restios are not showy and are generally rather difficult to identify. *C. mucronatum* ("bergriet" or "olifantsriet") is tall and frequently dominates the foreground, standing rigidly upright like a lot of thickish green sticks with brown tassels on top. *Elegia capensis* is attractive and is made much of in the gardens at Kirstenbosch. Somewhat resembling this, but tougher and often occurring in dense stands, is *Cannamois virgata* ("bergbamboes", "besemriet" or "perdehoef"). The name is derived from the Greek *kanna*, meaning "cane", and *virgata*, meaning "twiggy".

Mountain fynbos and other vegetation types

MOUNTAIN FYNBOS

Returning to the flowering plants of the open veld and the rocky slopes. Visitors from the northern hemisphere will meet many families well known to them. The daisy family (ASTERACEAE) is highly successful, and common everywhere. Our visitor would feel at home with such fine examples as the yellow, winter-flowering *Euryops* (meaning "large eye") *abrotanoides* ("geelmargriet") and *E. pectinatus* ("wolharpuisbos"). The genus *Senecio* is the world's largest. The name is derived from the Latin *senex*, meaning "without death", a reference to the persistent flower-heads. These two yellow daisies both lack ray florets (external petals), but are nevertheless striking plants.

The everlastings are different, the colourful, dry petals actually being the surrounding bracts (specialised upper leaves), not strictly part of the flower at all. Perhaps our best-known everlasting is *Helichrysum vestitum* ("strooiblommetjie", "tontelblom" or "witsewejaartjie"). *Vestitum* means "clothed", a reference to the white, woolly plants which flower best 2-3 years after a fire. The bracts are also white and when seen from a distance – particularly on Steenberg Plateau and in the Cape Point Nature Reserve – the massed plants resemble a large flock of sheep.

Daisies characteristically carry heads containing many flowers (hence the old family name COMPOSITAE), but in a few of our local genera the flowers in each head have become reduced to one only, surrounded by the usual bracts. This is the case with *Stoebe*, which includes the well-known grey "slangbos", and the purple *Corymbium*, another post-fire flowerer with long, narrow, parallel-veined leaves, more like that expected in a monocotyledon (bulb or grass-type plant).

The family PAPILIONOIDEAE, meaning sweet-pea-like, is well represented here, as it also is in Europe, popular members being the "keurboom" and "keurtjie" (*Virgilia oroboides* and *Podalyria calyptrata*), both sweet-scented forest margin trees; the large, blue-flowered *Psoralea* spp, "bloukeur, fonteinbos", which indicate moisture; the tall, often prickly, yellow-flowering *Aspalathus* spp; the strikingly beautiful silver pea (*Priestleya tomentosa*); and the remarkable orange nodding head, *Liparia splendens*, also known as the mountain dahlia or "skaamblom", a reference to the large, hanging flower-heads.

The local "bush tea" is made from the small, compact yellow-flowered bushes of *Cyclopia genistoides* (Spanish broom). These appear, sometimes in abundance, on the slopes and plains a few years after a fire. Unlike rooibos tea, made from the *Aspalathus linearis* found in the Clanwilliam district, cyclopia tea must include the flowers. Hence it is also known as "heuningtee", or honey tea.

The genus *Indigofera* contains a number of common species. The name means "bearer of indigo", referring to the blue dye extracted from an Eastern member. Two bushy, autumn-flowering species can be seen in the Kirstenbosch area. *I. cytisoides* (meaning "resembling cytisus"), another broom, though the flowers are mauve rather than yellow, is one of few shrubs that can live under pines. *I. filifolia* (meaning "thread-like leaves") is generally seen without any leaves – they fall off when young – so that photosynthesis is undertaken by the green stems and petioles (leaf stalks).

Indigoferas can be recognised by one particular

Wild dagga (*Leonotis leonurus*). (*Photo: D Hey*)

character. They carry hairs that are attached to the leaf or stem at their centres; each hair having two equal branches, arising at a common point. This feature can usually be seen with a 10x hand lens; do carry one in order to obtain maximum enjoyment from the flora!

Wild dagga, *Leonotis leonurus*, also known as Cape dagga, klipdagga, red dagga, wild hemp or minaret flower, is totally unrelated to real dagga, extremely unpleasant to smoke – according to the great amateur botanist Rudolph Marloth, author of the series *The Flora of South Africa* – and has little or no narcotic effect, although it has been used extensively as a medicine for skin diseases, including leprosy, piles (as a tincture), snakebite, paralysis and chest disease. The plant bears a very superficial resemblance to dagga, but can be readily distinguished by its simple leaves (true dagga has a highly divided leaf). Leonotis is up to 2 m tall, with orange-coloured flowers borne in whorls in midsummer, and grows on lower slopes, often near the sea. It also occurs on the contour path above Kirstenbosch. The name means "lion's ear", from the appearance of a single flower; "lion's tail" from the resemblance of the whole inflorescence.

We cannot leave the peas without mention of the scarlet-flowered *Sutherlandia frutescens*, a denizen of the hot, dry, rocky northern slopes. The inflated, green and brown, stalked, papery pods may occur at the same time as the flower. The wide distribution and striking appearance of this plant have led to many vernacular names – "kankerbos" (because of its use for treating internal cancers), "kiepiebos", "kipkippie", "klappers", turkey flower, "gansieblom", "hoenderbelletjies", "kalkoenbelletjies" (the name used by Van der Stel in 1685), "eendjiesblom" (because the pods resemble

"Kankerbos" (*Sutherlandia frutescens*). (*Photo: W P U Jackson*)

ducks and float on water), and "jantjiebarend" (said to be after the flaming red hair and beard of one Jan Barend).

The geranium family, GERANIACEAE, is known in Europe for its geraniums and erodiums (cranesbills and heronsbills, after the appearance of the fruit), but the gardener's "geraniums" are really pelargoniums (storkbills) and virtually all these are South African in origin. Pelargoniums have a Gondwanaland distribution and are not found wild in north-temperate zones. Although popular local plants, pelargoniums are not typical fynbos species, some even being succulent. They probably came to us from the north (e.g. Namaqualand). *Pelargonium cucullatum* (meaning "hooded") is an attractive and common local species. Somewhat similar is *P. betulinum* (after Betula, the European birch, because of the resemblance between the leaves), known as "kanferblaar" or "maagpynbossie", though this species tends to be lower and more straggly. *P. gibbosum*, the gouty pelargonium, is a strange, straggling species with swollen joints, found near the sea.

The lobelia is another garden favourite of South African origin. Its family, CAMPANULACEAE, is represented in the northern hemisphere by various bell-shaped flowers and it is better placed in a family of its own, LOBELIACEAE, thus separating lobelias from roellas, which in the Cape bear blue and white, bell-shaped, midsummer flowers. One particularly attractive lobelia is *Lobelia setacea.* The name honours Matthias de l'Obel, 1536-1616, Flemish nobleman, botanist and physician to King James I of England; *seta* is a bristle, referring to the short-haired anthers.

The borage or bugle family, BORAGINACEAE, also with predominantly blue flowers, is represented in both hemispheres, here particularly by the genus *Lobostemon. L. fruticosus* (meaning "shrubby") often has some pink flowers, perhaps more common when growing in the less acid shale than on sandstone. One common name is "agdaegeneesbos", because it was said to cure sores within eight days; another is "douwurmbos", because a concoction of the leaves was used to treat ringworm.

Members of the carrot family (APIACEAE, better known as UMBELLIFERAE), such as the European fennel and cow parsley, are easily recognisable, as should be the local blister bush, *Peucedanum galbanum.* The blister bush or "bergseldery" has large, shiny compound (pinnate) leaves and prominent greenish-yellow umbrella-shaped flowers which appear in summer. It is advisable for prospective mountaineers to learn about this common inhabitant of the damp slopes, for mere contact with the leaves or stem followed by sunlight can lead to blistering of the skin (phytophotodermatitis). Despite contrary opinions, no one is immune to it, though dogs are. It is an important food-plant for the larvae of *Papilio demodocus,* the Christmas butterfly. Other local members of the family look quite different, such as *Arctopus echinatus* (*arctopus* meaning "bear's paw", *echinatus* meaning "hedgehog" or "sea-urchin"), commonly known as the spiny bear's foot, which has large prickly leaves lying flat on the ground – "a nuisance to bare-footed people", as Thunberg punned in 1772. The unisexual flowers appear in midwinter in the centre of the leaf-rosette. "Pokkiesdoring" (pox-prickle) was a vernacular name, since, according to the historical Cape figure and army surgeon Dr James Barry, the exudation from the root of the plant was a sure cure for syphilis.

Blister bush (*Peucedanum galbanum*). (*Photo: National Botanical Institute*)

Blister bush in flower. (*Photo: National Botanical Institute*)

The rose family, ROSACEAE, is represented here by the large genus *Cliffortia,* which comes in all shapes and sizes but nowhere looks much like a rose. Its leaves are extremely variable but all carry stipules (i.e. small lateral appendages at the base). In fact, the saying goes, if you find a strange plant with stipulate leaves, call it a *Cliffortia* sp until proved otherwise! The damp upper slopes of Table Mountain viewed from, say, Rondebosch, show great patches of pale green which are composed of *C. odorata* (so-called although it is unscented).

Cape members of the gentian family are not blue-flowered; the best known has pink flowers followed by red berries. *Chironia baccifera* is named after the good centaur, Chiron; the specific name means "bearer of berries". This shrub of the lower slopes (e.g. around Kirstenbosch) has many vernacular names: Christmas berry (though the berries appear only around Easter); "aambeibossie", "barsbossie", "bitterbossie", "kalwerbossie", "lymbossie", "oondbossie", "perdebossie", "tandpynbossie", and more. The plant has served many medicinal purposes, most particularly as a remedy for piles and toothache; the ripe berry is edible.

Two families not found in the northern hemisphere are THYMELAEACEAE and SELAGINACEAE; both are common among mountain fynbos on the slopes. The flowers of the former are small but numerous and long-tubed. In *Gnidia* the florets are carried in terminal heads: in *Struthiola* they are axillary, up and down the stem. A feature of this family is that the long stems do not break – they tear. The purple, small-flowered, large-headed selagos are summer favourites; the flower-heads somewhat resemble the scabious. Incidentally, two *Scabiosa* spp also occur here, generally on rather damp slopes.

RENOSTERVELD

Renosterveld grows on the richer and less acidic clayey soil produced by the breakdown of Malmesbury shale and bears karoid, grassland and subtropical affinities. It contains few endemics and is often dominated by the renosterbos (*Elytropappus rhinocerotis*), a grey and unattractive shrub of the daisy family. Cape Town was built on shale, and today the best local example of renosterveld can be seen on Signal Hill, Lion's Rump and Little Lion's Head. Associated species include several *Rhus* spp ("taaibos"); *Passerina* spp (meaning "little sparrow"); *Felicia fruticosa* (wild aster or "bloublommetjie"); *Eriocephalus africanus* ("kapokbossie" or "wilderoosmaryn"); *Muraltia* spp; *Stoebe* spp ("slangbos"); *Metalasia muricata* ("witsteekblombos"); *Chrysanthemoides* spp ("bietou"); and several umbelliferous genera (family APIACEAE). *Cheilanthes* is a genus of ferns adapted to growing in dry shale; it may brown and dry up in summer, but re-greens with the rains.

In one small area we can still find the fascinating stapelia, now named *Orbea variegata*, or carrion-flower; it smells of rotten meat and attracts and is pollinated by flies. This was one of the first four local plants to be introduced to Europe, the other three being *Haemanthus coccineus* (April fool); *Cotyledon orbiculata* ("honde-oor"); and *Kniphofia uvaria* (red-hot poker). All were recorded by Johannes A Stapel in 1644. At one time the stapelia was quite common on shaley ledges, in the company of another succulent, *Euphorbia caputmedusae* (Medusa's head), but it is now much scarcer, probably because of the activity of collectors.

Between the lower shale and the upper sandstone, Lion's Head is composed of granite and conditions particularly suit the silver-tree. On this substrate the flora tends to be mixed, with the occasional renosterbos in evidence.

FLORA OF THE ROCK WALLS

Table Mountain sandstone tends to be fissured in two planes, producing countless more or less horizontal rocky ledges which possess their own characteristic floral communities. On the drier north and west faces the succulent red crassula is king of the summer, and perhaps one could say that the rich, yellow-flowering, grey-leaved *Euryops abrotanifolius* is king of the winter.

Red crassula (*Crassula coccinea*). (*Photo: W P U Jackson*)

Others of the daisy family are the succulent *Othonna arborescens* and *O. dentata*, and some white-flowered *Helichrysum* spp. Bright patches of mesems, or "vygies", such as *Lampranthus*, *Drosanthemum*, *Ruschia* and *Delosperma*, produce the most outstanding spring and early summer colour. Ericas are represented by *Erica plukenetii*, the red *E. abietina* (meaning "small pine-tree") and the pink, small-flowered *E. curvirostris*, together with some minor genera, especially *Blaeria* spp.

Coleonema album (klipbuchu or Cape may) is a rather special ledge-dweller, which produces a superb aromatic scent from its tiny leaves and its masses of little round white flowers. It is a relative of *Agathosma* of the citrus and buchu family. The woody stems are often bent, gnarled, thick and moss-covered and therefore look very old. *Cliffortia ruscifolia*, better known as the "climber's friend", has a similar woody stem that grows out of rocky crevices. It has such a firmly-based rooting system that rock-climbers can use the stem as a handhold or even as a roping belay. Its unfriendly aspect is its sharp, prickly leaves. It belongs to the rose family, yet has male and female flowers on different bushes. On Table Mountain itself only the tiny female flowers seem to exist, with their comparatively large, red, feathery stigmas; the plant is wind-pollinated. Could the ovules possibly develop new plants without being fertilised?

Scent

The marvellous fresh, aromatic smell of the fynbos is

very largely due to the glandular leaves of the citrus family, RUTACEAE. From elsewhere we obtain our oranges and lemons; here we have *Agathosma* (meaning "good scent"), *Diosma* (meaning "heavenly scent") and *Coleonema*, all commonly known as buchu, with various prefixes – klipbuchu, mountain buchu, bitter buchu, steenbok buchu, etc. The white-flowered, rock-dwelling *C. album* is also known as "aasbossie", from the habit of fishermen rubbing their hands on the leaves to rid them of the reek of red bait ("rooiaas").

COASTAL SCRUB

Comprising a mosaic of fynbos, strandveld and renosterveld, it occurs on rather deep, well-drained, compacted, sandy soil, limited in the Peninsula to a few narrow coastal strips that have not succumbed to invading wattles. It includes a lot of scrub-forest, often dense, with shrubby trees that remain dwarfed, such as the milkwood, *Sideroxylon inerme*; *Olea exasperata*, which is rather like a stunted Cape olive; and *O. capensis*, the bastard iron-wood with white flowers a few inches off the ground, almost inconceivably the same species as the giant forest specimens. *Cassine peragua*, the bastard saffron, also occurs in dwarf form, as does its cousin *C. maritima*. The *Maytenus oleoides* look-alike, *Pterocelastrus tricuspidatus* ("kershout"), bears bright orange three-cusped fruit (hence the specific name), while *Euclea racemosa*, the sea guarri, has a similar leaf shape.

In the more open areas several of the shrubs are succulent or semi-succulent, including *Euphorbia mauritanica* and *Aloe saponaria* (the soap aloe – actually used by early settlers for this and numerous medicinal purposes). Some of the shrubs are spiny, the *Lycium* spp, for example. Showy gazanias, vygies and annuals brighten up the area in spring, while the presence of several restios reminds us of the admixture of fynbos.

PARASITES

Total parasites have neither leaves nor chlorophyll and nothing green; they obtain all their nourishment from their hosts. The brightly flowered *Harveya* spp of the foxglove family (SCROPHULARIACEAE) are quite common during the summer months, the flower stems springing directly from the earth. *H. capensis* has large, white, crinkly flowers delicately tinged with pink; *H. purpurea* is deep pink; and *H. bolusii* is bright scarlet, the latter being found particularly on the lower table of Table Mountain. The name honours Dr W Harvey, treasurer-general of the Cape Colony and principal author of the early volumes of the *Flora Capensis*, and Harry Bolus, the stockbroker turned botanist who created the Bolus Herbarium. All harveyas turn black when bruised and were used as ink by the early settlers, hence their common name of "inkblom".

Closely related is the strange *Hyobanche sanguinea*, whose flower-heads, the colour of "ripe watermelon flesh" (Levyns), push upwards directly from the ground. Each red-hooded corolla tube ends in a hooked white projection, resembling a cat's claw or a tiny tortoise. Hence such common names as "katnael" and "skilpadblom", among others. The full name, given by Linnaeus, appears to mean "bloody strangled pig".

The semi-parasitic, insectivorous sundews (*Drosera* spp) have leaves with glandular hairs that secrete a proteolytic enzyme capable of breaking down the tissues of insects that get caught in the sticky secretion. The metabolites of the insect cells are absorbed and nourish the plants. The leaves, often held in a basal rosette, are reddish-green and glisten in the sunlight. The tiny utricularias need even more moisture as they must maintain the minute underwater bladders on their stems with which they can catch and digest microscopic aqueous insects. Their little yellow and mauve flowers are often seen above their damp mossy bed. The name is derived from the Latin *utriculus*, meaning "a small leather bag" – a reference to the tiny bladders.

Sundew (*Drosera* sp). (*Photo: W P U Jackson*)

Plants that are parasitic on stems live entirely above the ground, their suckers entering the stem-tissue of the host. One such is the yellowish twiner *Cassytha ciliolata* (devil's tresses), which can be found in the upper parts of Kirstenbosch. It belongs to the family LAURACEAE, the laurel family, as also does the stinkwood tree (*Ocotea bullata*). Mistletoe (*Viscum* spp) is another aerial parasite, rather sparsely scattered around the Peninsula, attaching mainly to *Rhus* spp.

They flower at any time, hence the common name "altydbos". *S. glutinosa*, with larger, sticky flowers, is rather rare in the northern parts. Look out for it in Nursery Ravine, quite near the path.

Devil's tresses (*Cassytha ciliolata*). (*Photo: W P U Jackson*)

Members of the PENAEA family have a squarish look – their parts are in fours, stems often four-angled, leaves closely packed, in alternately opposite pairs. The type-genus, *Penaea*, is common but not attractive, while *Saltera sarcocolla* is a striking shrub, the inflorescence consisting of a few congested brilliant pink flowers with extruded stamens, surrounded by sticky leaves or bracts. The broad stem-leaves are frequently coated with a waxy white bloom. The name honours Captain T M Salter, an officer in the Royal Navy who became the co-author of the well-known *Flora of the Cape Peninsula*, and the specific epithet means "gluey flesh".

Brachysiphon fucatus is a beautiful but rare shrub confined to rocky parts of the northern mountains, producing deep pink flowers in midwinter. It can be seen on the summit of Grootkop and at the top of the path up Nursery Ravine.

"Rooistompie" (*Brunia stokoei*). (*Photo: W P U Jackson*)

ENDEMIC FAMILIES

Families not represented in the northern hemisphere, in fact occurring only within the fynbos biome, include the distinctive BRUNIACEAE and PENAEACEAE. Brunias and berzelias are easy to recognise from their persistent inflorescences of crowded, spherical flower-heads. When in full flower each little head somewhat resembles a disintegrating golf ball. *Brunia nodiflora* ("stompies") can be dotted about in the veld throughout the mountain chain.

Berzelias are very similar (they have a single-chambered ovary), but they may often be distinguished in the field by a red waxy thickening of the flower-stems. The purpose of this appendage is unknown. Berzelias all indicate dampness – a clump of *Berzelia lanuginosa* ("kolkol") being a clear danger signal. This berzelia has smaller flower-heads (about 5 mm in diameter) than the others, and needle-like leaves.

The staavias are rather different; the flowers are aggregated into a disc rather than a ball and are surrounded by brilliant white petal-like leaves, closely resembling a daisy. *Staavia radiata* is common and widespread, with small flower-heads about 4 mm wide.

"Kolkol" (*Berzelia lanuginosa*). (*Photo: W P U Jackson*)

Saltera sarcocolla. (*Photo: W P U Jackson*)

INVADERS FROM THE KAROO

Some plants seem out of place in the fynbos – especially the succulents, which were primarily adapted to a hot, dry climate. The CRASSULA family for instance has done so well that it has produced species that are found only in fynbos. The outstandingly fine summer-flowering red crassula, *Crassula coccinea*, is a case in point, but its favourite habitat is among well-drained rocks that are dry in summer. *C. rupestris* (meaning "of the rocks") is common among the dry koppies of the Karoo but is also found among rocks on the Cape Peninsula. The large, thick, round leaves of *Cotyledon orbiculata* ("varkoor" or "honde-oor") are common on sandy soils near the sea. The fleshy internal part of the leaf has been used to treat corns, warts and open sores.

One is surprised to find the related *Tylecodon paniculata*, the well-known "botterboom" of the Karoo, growing happily below cliffs on our west coast. In the same sort of habitat we can see the succulent, much-branched *Euphorbia mauritanica* ("geelmelkbos"), while the strange *E. caput-medusae* ("noordpol" or "vingerpol") also occurs on the lower western slopes. If you are lucky, you may find the stapeliad *Orbea variegata* (syn. *Stapelia variegata*), commonly known as the carrion-flower, on lower granite slopes, flowering in autumn and often close to *Euphorbia caput-medusae*. It is the only stapeliad in this area.

We are indeed enriched by our succulent mesems, or "vygies" (family MESEMBRYANTHEMACEAE, meaning "middle-of-the-day flowers"). All have brilliantly coloured flowers, forming bright patches of red, pink, yellow or orange that cover rocky surfaces, especially in spring and summer. The most common local members are *Lampranthus*, *Ruschia*, and *Carpobrotus*, the sour fig or Hottentot's fig. After they have dried on the plant, the fruit of the latter can be eaten or made into jam, while the flesh of the leaf will reduce the pain and swelling of bluebottle stings. The genus *Erepsia*, while not as showy as some of the others, has the distinction that the flowers stay open in all weathers. The popular annual "bokbaaivygie" *(Dorotheanthus bellidiformis)* flowers in white sand at Kommetjie and Camps Bay. The name means "Dorothy-flower" after the mother of Dr Gustav Schwantes, a German botanist and expert on succulents, and the specific epithet means "resembling Bellis", the English daisy – which is not very appropriate.

ANNUALS

Annuals are poorly represented in the fynbos. The wonderful sheets of colour on our hillsides are composed largely of perennials, ericas, leucadendrons, vygies and perennial daisies. Some annual senecios with purple and yellow flowers grow near the sea and on the lower slopes. *Senecio arenarius* (meaning "of sandy places") is known as the "hongerblom" because it is used as a tea for indigestion. The leaves of *S. elegans*, the wild cineraria, can be cooked and eaten as a spinach. The feature that differentiates these two species is the hairiness of the involucral bracts in *S. arenarius*. The only annual seen in any mass by the average Capetonian is *Dimorphotheca pluvialis*, the rain daisy, or snow daisy, that appears in early spring above De Waal Drive and on the slopes of Signal Hill.

GEOPHYTIC FAMILIES

Geophytes are plants that spend part of the year entirely underground. These are mostly bulbous, cormous, tuberous or fibrous-rooted monocotyledons, but the widespread and showy *Oxalis* genus is included here. Indeed, the latter are bulbous plants, with contractile roots that carry the bulb deep into the ground and with a built-in clock that tells them when to sprout, even when deprived of moisture, light and soil. It is a large genus with many species that are difficult to identify. Two species that have adapted to our gardens we know well – *O. purpurea*, with large pink flowers that like to come up in lawns, and *O. pes-caprae* (meaning "goat's foot", from the shape of the leaf), a long-stalked, yellow-flowering pest, though popular as a pot-plant in the West Indies. *O. incarnata* we have already mentioned, and we should just refer to the superb, low-lying, ground-flowering *O. luteola*, with its rich yellow flowers, that appear on the lower hills in autumn. Geophytes make up an unusually large proportion of fynbos vegetation – and this goes also for renosterveld and strandveld.

Orchids are advanced plants with complicated floral structures. One feature is that all the pollen grains in each lobe of the anther cohere and carry a sticky gland, the viscidium, so that when an insect touches it the whole structure comes away sticking to the pollinator, which takes it along to the next flower – nature's form of precision gambling. *Disa* is the best-known genus of our orchids. It was named by the Swedish botanist Petrus Bergius in 1767, after a legendary Swedish queen. The king was looking for a wife, but she had to enter his court neither naked nor clothed, neither riding nor driving. So Disa arrived clad in a fishing-net, one foot on a sledge and one leg over a donkey. The triumphal entry of Disa into the royal court is still celebrated annually in Uppsala, the home of Linnaeus.

The magnificent red disa (*D. uniflora*), also known as the pride of Table Mountain, flower of the gods, and the emblem of the Mountain Club of South Africa, is the largest of all South African orchids and some people think the finest terrestrial orchid in the world. Good specimens can be over 50 cm tall and carry five or more blooms, each with a spread of 12 cm. They grow on the banks of streams and damp rock-faces and flower in February-March – at one time in such profusion that Sir Percy Fitzpatrick, politician and author of *Jock of the Bushveld*, reported that they covered the area around the present Woodhead reservoir "like a scarlet blanket". The specific name *uniflora* is unfortunate, since multi-flowered stems are common; apparently its author, Bergius – a pupil of Linnaeus – named from pressed, single-flowered specimens. Linnaeus the younger converted this to *grandiflora*, but under modern rules the original name has to stand.

The blue disa (*Herschelia graminifolia* – now, strictly, *Herschelianthe*), is just as beautiful, but much smaller. Isolated stems with up to eight blooms are quite common, flowering in late summer among restios in the open veld above 300 m on Table Mountain. The name honours Sir John Herschel, the astronomer, who also cultivated orchids; the specific epithet indicates the grasslike leaves. In the southern Peninsula the blue disa is replaced by the very similar early purple disa (*H. purpurascens*), which flowers in November.

The mauve or drip disa (*Disa longicornu*) is a popular Table Mountain rarity, blooming for Christmas only on damp, south-facing rock-bands that receive a constant drip of water – at least until after flowering. The unusual lilac colour, the large hood coursed by green veins and the long horns or spurs make an unforgettable picture.

The cluster disa (*D. ferruginea* – meaning "ferrous" or "rusty"), is an attractive, many-flowered orchid, its rufosity blazing among the rocks to greet visitors as they emerge from the cable station on the top of Table Mountain during the late summer months.

Flowers in the genus *Satyrium* are usually several in a dense spike and each carries two spurs, hence the fanciful resemblance to the satyr – half-man, half-goat (or originally half-man, half-horse in Greek mythology). *S. carneum* (meaning "fleshy") is a large, pink, spring-flowering orchid up to 1 m tall. It grows on low, sandy slopes, generally at or near the coast, southwards from Llandudno, and is common on the east side of the Karbonkelberg. The similar but rather smaller, orange-coloured *S. coriifolium* (meaning "leathery leaf") occurs

Ewwa-trewwa (*Satyrium* sp). (*Photo: W P U Jackson*)

more on inland plains. Both are commonly known as ewwa-trewwa.

Though there are many more orchids in our area we will conclude with the winter-flowering *Disperis capensis*, the granny bonnet or "moederkappie". Each side sepal carries a little pouch, *disperis* (meaning "two-pouched"), while the magenta side-petals project beyond the hooded upper sepal to produce the effect of a bonnet or "kappie". Flowers are usually single but easily found, scattered around in the veld on plains and mountain-tops.

The iris family is well represented by many splendid genera and species. This is perhaps my own favourite family, ranging from the tall, fibrous-rooted, brilliantly blue aristeas and multicoloured *Witsenia maura* (bokmakierie's tail) to the dainty and charming cormous gladioli and moraeas. We have already met *Watsonia borbonica* (*W. pyramidata*), a prolific, tall, pink, post-fire flowerer. When intermixed with one of the tall blue *Aristea* spp on the lower slopes, usually around Christmas, it presents an unforgettable sight. The endemic *W. tabularis* (meaning "of Table Mountain") occurs on the upper slopes, flowers slightly later than *W. borbonica* and is an unusual salmon to coral-pink colour. South of Table Mountain the colour changes to a brilliant scarlet (var. *concolor*). This interesting species is also distinguished by possessing stem-sheathing leaves that form a water-retaining cup. This is sometimes useful on a hot day; have a look for yourself. Another post-fire flower grows on the southern flats – the bright orange *W. zeyheri* (syn. *W. comptonii*, after a late director of the National Botanic Gardens). The generic name honours Sir W Watson, MD, a London apothecary.

Painted lady (*Gladiolus carneus*). (*Photo: Percy Sargeant Memorial Collection, National Botanical Institute*)

Beware of being tripped by the tough, thin, elongated leaves of one of the two tall bobartias ("besembiesie"). *Bobartia indica* (Linnaeus mistakenly believed the plant had come from India) possesses a cylindrical stem and leaves; in *B. gladiata* both are flattened. Both species carry dense clusters of yellow flowers at the end of a long stem.

Gladiolus carneus (painted lady or "bergpypie") is probably the best known gladiolus. The flower is cream and pink with red or purplish spade markings (honey-guides) on the three lower tepals. It is quite common on the slopes, being particularly robust after a fire, and carries several almost horizontal branches of flowers. *G. brevifolius* is another common "pypie" of the slopes with a long, summer-flowering season and small spikes of pale pink flowers with yellowish spade markings. There are a few very short leaves on the stem (hence the name); the single long basal leaf is not present during the flowering season. This so-called hysteranthous habit is a common adaptation among midsummer-flowering monocotyledons, enabling the energy-producing leaves to be active in the damp/cool season and protected from desiccation during the hot/dry season. Gladiolus means "little sword" – a reference to the leaves.

Ixias have slender stems, narrow leaves and spikes of regular (radially symmetrical) flowers arranged spirally. They are known generally as corn-lilies, wand flowers, or "kalossies". *Ixia polystachya* (meaning "many spikes") frequents the damp eastern slopes of Table Mountain in November in several different colour-forms. The commonest is white, pink or mauve with a deep central stain, but a blue variety occurs in the Constantia ridge area, while the yellow form can be seen around the summit of Skeleton Gorge. In contrast the yellow ixia (*I. dubia*) flowers in dry, open spaces on the western side. The flowers are bright orange with a dark central stain, previously incorrectly known as *I. maculata*. Ixia is probably derived from the Greek for bird-lime, on account of the viscous sap.

The blue and white, low-flowering babianas, "bobbejaantjies", are always appreciated and can be recognised from their pleated, hairy, fan-wise leaves. The name is a latinised version of the Afrikaans for baboon, since this creature is fond of babiana corms. Men have also found nourishment from these, and also the corms of ixias and watsonias; but beware of some moraeas and all homerias, "tulpe", which also belong to the iris family.

The corm of the sweet-scented *Moraea fugax* ("uintjie") is edible; it was previously known as *M. edulis*. The flowers are white, yellow or blue, usually accompanied by a single long leaf. The blue *M. tripetala* has a long flowering season, a wide distribution in the

Gladiolus caryophyllaceus. (Photo: W P U Jackson)

(pronounced something like "K-nip-hofia"), and our *K. uvaria* is the parent of most of the world's garden varieties. It grows naturally in swampy places, mainly in the south.

The bulbous chincherinchee, or "chink", needs no description. The popular large white species is *Ornithogalum thyrsoides*, and it occurs in damp and shady places. The generic name means "bird's milk", and a *thyrsus* was originally a staff surmounted by a pine cone, carried by Bacchus. The smaller bright yellow chink is found on the west coast. The family also contains the bright yellow bulbines and bulbinellas, "katstert".

Aloe succotrina is a sprawling plant with red and greenish-yellow flowers in midwinter. It occurs in a large patch among boulders deep in Newlands forest – a very odd place for a succulent aloe. Further south it grows on open, rocky ledges. The name means "yellow (citron-coloured) sap".

Peninsula, and, at first glance, only three narrow tepals. It is thus easy to recognise. The name *Moraea* was originally given in honour of an English Dr Moore, and spelt Morea, but later appropriated by Linnaeus to commemorate either his wife, Sara Elizabeth Moraea, or his father-in-law, Dr Johan Moraeus.

There is only one amaryllis and that is our beloved pink *Amaryllis belladonna* (meaning "beautiful lady"), the March lily. (The florist's "amaryllis" is a hippeastrum.) In autumn the buds pop out of bare ground and rapidly flower, being particularly prolific after a fire, quite often accompanied by the related red *Haemanthus* spp, the April fool, also called "misryblom".

The same family contains the world-famous nerines, and especially our own *Nerine sarniensis*. The backward-curving, crimson-red tepals appear to be sprinkled with gold dust when viewed in the sun. This phenomenon is caused by the refraction of white light in the superficial layer of cells, breaking it up into colours rather as a diamond does. The flowers appear in autumn, on rocky ledges and shady slopes. The specific name means "from the Channel Isles". Nerine bulbs collected in the Cape were washed up on the shore of Guernsey from a shipwreck in 1659, grew and flowered and were for 100 years believed to have originated there; even today they are called Guernsey lilies.

The lily family contains some well-known Cape members, though no true lilies. *Agapanthus africanus* is a fine, summer-flowering species of the slopes with deep blue, spherical inflorescences. The name means "love-flower"; why the Frenchman L'Héritier chose it is not clear. Red-hot pokers are correctly called *Kniphofia*

Special communities

INDIGENOUS FORESTS

Indigenous forests occur in the Newlands-Cecilia area, Orange Kloof, Tokai (much here is exotic), Kalk Bay mountains (two old relic forests), above Noordhoek and in kloofs on the Camps Bay side (the north-western slopes) of Table Mountain. Above Hout Bay, so named because of its forests, barely an indigenous tree remains. Odd patches of forest trees occur elsewhere in damp mountain recesses and in the shelter of clusters of giant boulders. Our forest trees are almost all of "Afromontane" origin – they have migrated southwards from the mountainous areas of east and central Africa, and then westwards via Knysna. On the way species have dropped out; thus we have fewer species than Knysna but still far more than are found in the forests of the northern hemisphere. Probably through the climatic variations in the Pleistocene, damper eras allowed migration of forest trees while drier eras saw their recession.

Our forests were probably once far wider in extent, but we have no certain facts. There is little doubt that there were many larger trees in the forests when Van Riebeeck landed – he recorded in his diary how rapidly the best timber was cut out. Today, the Newlands forest (at Kirstenbosch) boasts a single decent-sized stinkwood (*Ocotea bullata*), which, presumably, the early foresters overlooked. The younger stinkwood trees can be recognised by leaves which carry a few obvious swellings (*bullae*, hence the species name) at their base. These bullae are believed to be the homes of mites. Smaller, less obvious bullae are present in *Chionanthus* ("baster-fynblaarysterhout"), *Kiggelaria* (wild peach), and *Canthium inerme* (meaning "turkey-berry").

The forests still contain some sizeable trees of that other fine cabinet wood *Podocarpus latifolius* (the real, broad-leaved yellowwood); *P. falcatus* is presumed to have fallen out around Swellendam in the southwards migration. Other important trees are the ironwoods (*Olea capensis* and *Chionanthus foveolatus*); the assegai (*Curtisia dentata* – meaning "toothed leaves"); the bastard saffron (*Cassine peragua*), with orangy-red bark; and the "boekenhout" (*Rapanea melanophloeos* – meaning "black wood"), with rather long dark leaves. *Olinia ventosa* (the hard pear) produces some of the largest trees. It is recognisable by its bark which resembles the scales of a fish or the shedding of tears, while the crushed leaves exude the scent of bitter almonds.

We have already mentioned the wild almond (*Brabejum stellatifolium*). The tree fuchsia (*Halleria lucida*) carries small, tubular, orange flowers and berries on its trunk and main stems; the "rooi-els" (*Cunonia capensis*) bears fine spikes resembling white candles at Easter, large compound leaves, and strange central butterspoon appendages (actually stipules). The "keurboom" (*Virgilia oroboides* – meaning "of the mountain") is a "nurse" or precursor tree, growing ahead of an advancing or recovering forest. The wild peach (*Kiggelaria africana*) is rather similar. It grows rapidly, is frequently found in gardens and is sown by bird droppings. In some years this tree is almost totally defoliated by the brownish caterpillar of the *Acrea horta* butterfly, but is seldom killed by it. The cuckoos are the only birds that feed on these caterpillars. The wild olive (*Olea europaea*) is another well-known tree of forest margins and coastal scrub. Long called *O. africana*, it is now believed to be a variety of the European olive of commercial and biblical fame.

Most, if not all, trees that bear berries attract birds, which carry the enclosed seeds away and so allow for distant dispersal of the species. These include the olives (*Kiggelaria*), stinkwood and boekenhout. The fruit of the latter is also eaten by baboons.

The common names of most of our trees were given by the early foresters who compared their wood to that of well-known European trees. The boekenhout, for instance, in no other way resembles a beech.

THE FOREST FLOOR

Our forests have not developed an infrastructure comparable, for instance, to that of the great karri forests of Western Australia. Perhaps the canopy is too dense. *Oxalis incarnata* is prominent in places, forming a pinky-white carpet. Restios are not seen, but the sedge (*Schoenoxiphium lanceum*) is a common forest floor covering, with long, grasslike, sharp-edged leaves (hence the *lanceum*). The April fool (*Haemanthus* – meaning "blood-flower") is the outstanding inhabitant of the Spes Bona forest in the Kalk Bay mountains. The white-spathed Cape arum lily is another monocotyledon, or bulb-type plant, found mostly in damp kloofs, where it may flower until Christmas. Its proper name is *Zantedeschia aethiopica*, after an Italian botanist – there are no true arums in South Africa. *Knowltonia* spp ("katjiedrieblaar"), which have green flowers and tripartite leaves, are collected locally for medicinal purposes. They are one of our few examples of the buttercup (*Ranunculus*) family.

Flowering in May is the bright red *Chasmanthe aethiopica* ("aethiopica" simply referred to Africa south of the known world, i.e. Egypt); closely related is *C. floribunda*, which grows on ledges. Both are known as "suurkanolpypie". Brambles (*Rubus* spp) are unpleasant forest aliens, but a more welcome foreigner is *R. cedronella*, which has a delightful scent supposed to resemble that of cedar wood; hence the name. We do not find a large array of climbers, but two big ones are *Rhoicissus tomentosa* (wild grape) and *Secamone alpini* (monkey-rope). Ground scramblers of the asparagus group are common, especially the non-prickly *Myrsiphyllum scandens*.

FERNS

Ferns are not solely forest-dwellers, although largely so. Apart from the heat-resistant *Cheilanthes*, several ferns can grow in any permanently damp spot, between rocks, in caves, or on stream banks. If you visit the Elephant's Eye cave on Constantiaberg you will be astonished to find *Blechnum* and *Gleichenia* growing in almost total

The young fronds of *Blechnum*. (*Photo: W P U Jackson*)

darkness. Bracken (*Pteridium aquilinum*) is an almost universal pest, a post-fire invader, but it does have the grace to die down for the winter. The tree fern (*Cyathea capensis* – syn. *Alsophila capensis*) is confined to densely wooded kloofs. Its "trunk" is porous, not woody, and is composed of old leaf-bases. At the apex of this trunk can be seen strange, fern-like branched appendages known as aphlebiae (meaning "without veins"). This fern can be seen at the Cecilia waterfall.

Two forest ferns sometimes called mother-in-law's tongue bear simple, undivided tongue-like fronds and grow on damp rocks or under trees. The only singly divided (pinnate) ferns are *Blechnum* spp, also distinguished by having fertile fronds (thin and dark), instead of the common sterile fronds. The young fronds of some species display beautiful soft shades of pink and brown. Other fern fronds are at least doubly divided (bipinnate) and quite difficult to identify. One of these is the maidenhair fern (*Adiantum*), of which we have three species. The name is derived from Greek and means "incapable of being wetted" – during a shower, the fronds shed water and remain dry. Most people will recognise these from their delicate appearance and the triangular shape of the smallest segments.

Ferns have a branched vascular system and propagate by means of spores. These are aggregated together in clumps called sori, which can be seen underneath the fronds in the right season. The disposition of these sori are important distinguishing features and worth looking for – just turn the frond over and examine its underside.

LICHENS

The first living things you see on Table Mountain may be lichens. Sandstone is white; the massive brick-red sandstone blocks visible for miles are coloured either by iron ore or by the lichen *Umbilicaria*. Closer inspection yields sulphur yellow (*Xanthoria*), brilliant green, black and other colours. They are often so dry and flat on the rocks that they are difficult to distinguish from plain mineral pigment. In damper conditions feathery lichens hang from trees, resembling the Spanish moss we have in our gardens (this is a bromeliad), and these have indeed been collected for the extraction of their pigments.

A lichen is a fascinating combination of an alga and a fungus, the resulting symbiotic organism being very different from either the alga or the fungus growing alone. Lichens do not need organic food as the algal part photosynthesises; they remain alive when completely desiccated, and they absorb most of their mineral requirements from the air and rain. Because of this they are highly susceptible to airborne toxins and are used as a sensitive indicator of air pollution.

Lichens. (*Photo: W P U Jackson*)

MOSSES AND LIVERWORTS

These small, lower forms of green plants are very common on damp surfaces. Collectively known as bryophytes, they do not possess a vascular system and therefore have to obtain their water and food by a process of diffusion from cell to cell. The moss plant (gametophyte) consists of a stem and leaves. From the basal moss plant one often finds stalked structures with capsular heads arising. These are the sporophytes that produce spores by an asexual process. Sphagnum is one moss that is particularly prized for its spongy, water-holding properties.

Liverworts can be distinguished from mosses by being flatter or cup-shaped, and often a greyish-green colour.

LARGE FUNGI – MUSHROOMS AND TOADSTOOLS

In autumn, large numbers of different fungi appear on the ground and on rotting tree stumps. The visible mushroom is the fruiting-body arising from a permanent underground meshwork, the mycelium, which is almost always saprophytic, i.e. living on rotting plant material. It is without chlorophyll (never green) and hence unable to photosynthesise for itself. There are several different "puff-balls" which remain entirely spherical and are of little interest to the average person. There are also a large number of different "bracket" fungi, forming shelf-like projections without stems on dead stumps or living trees. The sulphur-shelf, the name indicating its colour, is actually pathogenic on live oaks and bluegums. It is edible when young, when it is known as "chicken of the woods".

Most of our edible fungi have been imported with trees from abroad. The popular, edible pine-ring (*Lactarius deliciosus*) is a bright orange colour, turning partly green with age. It is illustrated in the frescoes of Hercu-

Fly agaric (*Amanita muscaria*). (*Photo: D Hey*)

"Belskruie" (*Osmitopsis asteriscoides*). (*Photo: D Hey*)

laneum and Pompeii. The purple-stemmed *Russula* is extremely common under pines, is variable in colour and has a peppery flavour.

The amanitas are interesting. *Amanita phalloides* resembles the field mushroom but is extremely poisonous; so is *A. pantherina*, which has a brownish cap and resembles the pleasantly edible *A. rubescens*, the pink-tinged blusher. Some amanitas, unless old, carry white flecks on their caps; particularly prominent is *A. muscaria* (fly agaric), the fairy toadstool with its brilliant red and white colours. Much used in ancient times for killing flies, it is poisonous and hallucinogenic – but no one seems to know how poisonous or how hallucinogenic the Cape specimens are, since their chemical composition and hence their properties differ in different parts of the world. All occur under pines.

With most mushrooms the spores are held in the gills, but in the boletus group the gills are replaced by tubes so that the underside of the cup resembles a sponge. No member of this family is poisonous, and one, the cep (*Boletus edulis*), is an epicurean delight. Growing under pines, it appears after the first good rains of the year and may be the size of a soup-plate. The Romans called it *suillis* and used it for both culinary and medicinal purposes – e.g. for healing dog-bites and preventing freckles. The poplar boletus (*Leccinum duriusculum*) is also good eating.

Beware of any completely white or off-white mushroom – the field mushroom is never all-white. Beware also of the panther – it can resemble the blusher quite closely. Compare any unknown mushroom carefully to the description and illustration in a reliable book.

WATER-LOVERS

Many flat areas on the plateaux are marshy in winter and may retain this character throughout summer if they are sufficiently enveloped in south-easterly cloud. Other permanently damp areas occur around the stream banks. These habitats attract certain plants, including the large, white-flowered daisy with the resounding name *Osmitopsis asteriscoides* ("belskruie", meaning "Bellis-herb", Bellis being the European daisy); *Berzelia lanuginosa* and, in the southern parts, *B. abrotanoides*. Frequently associated are the dull, red-flowered, curved-tubed *Erica curviflora*, the yellow-flowered *Gnidia oppositifolia*, the tall, blue-flowering *Psoralea aphylla* (meaning "without leaves", actually with very small leaves) and the "bloukeurboom" (*P. pinnata*). Various species of the little insectivorous *Drosera* (sundew) sparkle in areas where the ground is sufficiently open, as on the sides of a damp path or stream, while the tiny *Utricularia* abounds where the moisture is perennial.

In the same habitat occur two members of the gentian family – the charming, feathery, yellow-flowered *Villarsia* and the less common, purplish-pink *Chironia jasminoides*. In stream-beds and along their margins the tall *Prionium serratum* ("palmiet"), and the bamboo-like grass *Phragmites communis* predominate, occurring together with several other rushes (*Juncus* spp) and sedges (fam. CYPERACEAE) as well as the tree erica (*Erica caffra*). The king of stream-beds is the red disa.

Some damp plains are covered with restios. Many are thatching- reeds, and include the easily recognised, tall, thick-stemmed *Chondropetalum mucronatum*. One erica associated with restios in the southern parts is *Erica bruniades*. Two forest trees always found close to water are the "rooi-els" (*Cunonia capensis*), and the "waterhout" or "without" (*Ilex mitis*), with its whitish bark and dark, pointed leaves.

Pink senecios after a fire.

AT THE SEASHORE

Salt-tolerant, sand-dwelling communities include the coastal scrub, mentioned on page 56. Coastal sand-dunes are prone to wind erosion because of a natural deficiency of sand-binding grasses, which led to the introduction of some of the Australian acacias. Indigenous shrubs that help to bind dunes permanently include the waxberry (*Myrica cordifolia*), *Passerina* spp, *Metalasia muricata* and some restios. At a later stage some very fine flowering species appear, such as the daisies, e.g. *Senecio arenarius* ("hongerblom"); the vygies, e.g. *Dorotheanthus bellidiformis* ("bokbaaivygie"); *Lampranthus* spp; the large, brilliant-yellow *Conicosia*; the sea lavender or statice (*Limonium* spp); brown salvias; the odd little *Zaluzianskya villosa* (drumsticks); the shining *Orphium frutescens* (named after Orpheus, the miraculous flute-player); the brilliant-red, cormous *Anomalesia cunonia*; and many more. One of the best places to see a good variety of these colourful coastal flowers in spring is at Pegram's Point in the Cape of Good Hope Nature Reserve.

MARINE ALGAE

Approximately 200 species of seaweeds have been recorded on the Cape Peninsula. Of these, 100 species may be found on the shores of the Cape Point Nature Reserve. Three beaches are easily reached by visitors – Olifantsbos and Platboom on the western side, and Buffels Bay on the False Bay side.

Except for a few rare exceptions, all seaweeds must have firm surfaces for attachment. Sandy beaches are therefore seldom good places to look for seaweeds, although species which grow sub-tidally are often dislodged by strong wave action during storms and cast up on the beach. Although seaweeds are classed as either green, red or brown, depending on basic pigmentation, they have a wide colour range – all shades of green, pale pink to magenta, and beige to deepest black.

At Olifantsbos there are large outcrops of shale on the Cape Town side of the beach. On this shale on the higher parts of the shore, deep, smooth-sided rock-pools have been gouged out by years of wave action. The pools will only be reached by waves during high tides and are characterised by a deep fringe of the green seaweed *Enteromorpha* at their surfaces. A terrestrial botanist would term this species a weed, but algologists prefer to call it an opportunist – able to survive a wide range of temperatures and salinities. Deeper in these rock-pools other species of green, brown and red algae will be found, but not in great variety. At low tide, when lower regions of the shore are exposed, many more species of red and brown seaweed can be found. The common kelp

Common kelp off Cape Point. (*Photo: Paul Alberts*)

(*Ecklonia maxima*) is abundant in the breaker zone beyond the low-tide line, and large quantities are dislodged and washed up on the beaches after a storm. This is collected, dried and milled for sale as fertiliser.

Platboom Beach lies just below the car park, and outcrops of Table Mountain sandstone run out to the sea on either side. Here, the rocky surfaces afford possibly the best opportunity for seaweed study. In many places the sandstone has become fissured by wave action, but rock platforms with many shallow rock-pools may be found. Green sea lettuce (*Ulva* spp) is very common in these rock-pools, often growing quite large where it has not been grazed by animals or abraded by sand. Few species of seaweed can tolerate long exposure to the air, but a common exception is the red seaweed *Porphyra*. It actually appears brownish in colour and has a very thin, delicate-looking appearance which belies its actual toughness.

Buffels Bay has fewer species than either Olifantsbos or Platboom. This is because sandstone boulders, which may be moved by strong wave action and thus discourage algal growth, overlie the rock-base. Many of the boulders have patches of green algal turf. This is usually composed of hundreds of very short, tightly packed plants of the *Enteromorpha* spp. Algae are discrete and do not generally present the lavish displays found in flowering plants. The algal plants may take the form of pink or red patches, miniature brown balloons, green string, chains of green beads, pink or brown feathers, or green velvet fingers, to name but a few. Rarely does a species occur in such obvious numbers as the kelps. The rich beauty of colour and form is only revealed to the interested researcher.

The local kelp is well known for its use as a fertiliser and plant growth stimulant. It is also the source of alginic acid, used commercially as a stabiliser in foodstuffs. Red algae yield agar, allowing us a big economic saving from having to import this valuable substance, used mainly in the pharmaceutical trade.

INVASIVE ALIEN VEGETATION

The early settlers introduced European trees to the Cape, partly for sentimental reasons and partly because they felt that the comparatively shrubby and tree-poor veld needed trees. Trees were also introduced for practical purposes such as timber production, for hedges and windbreaks, for tanning, and for stabilising driftsands. Of the broad-leaved species the English oak (*Quercus robur*) was the most successful. The cluster pine was introduced by the Huguenots late in the 17th century, but aliens on a large scale appeared only after 1810. In the middle of the 19th century, both Baron Carl von Ludwig and the newly established Cape Town botanic garden brought in wattles, stinkbeans, myrtle, hakeas and other invasive species for "ornamental" purposes! In the 1860s, J C Brown, official botanist of the Cape Colony, urged the public to plant as many trees as possible of any kind, and distributed seeds freely.

This policy was continued after 1875 by J S Lister, superintendent of plantations, who established plantations of cluster pines all over the place, most of which were later abandoned. He also distributed pine seeds to mountaineers to throw around on the mountain! Lister encouraged the growth of Port Jackson wattle for commercial use in tanning and for export of the bark. These stands were also abandoned later and were then of use only for firewood.

Why are the invasive trees so successful? There are many possible reasons for this, not all fully understood and, of course, not all applying to every species.

Of all factors a lack of native predators – especially seed predators – is probably by far the most important, and if alien species are able to take advantage of open spaces after fires or other disturbances, it may well be the absence of their natural predators that allows this to happen. Hence one is ideologically swayed towards control by biological methods – i.e., the introduction of natural predators and diseases such as the root fungus.

Control of invasive species

Manual removal of plants by various means is very effective, but also expensive for heavy infestations. Nevertheless, it is the most important method employed by the many public "hack" groups – in some places with considerable success, but at enormous cost in terms of man-hours and dedication. Let us salute these consistent hackers!

Manual methods include hand-pulling of small plants, chopping and slashing, sawing, mattocking and digging out, and ringbarking. Mechanical methods include use of a brush-cutter and a mower, or ploughing out the roots. Herbicides may be used as sprays, or for stem, stump, coppice, seed or soil application. Burning may kill seedlings or mature plants, or reduce the seed store. Various combinations are generally employed, such as chain-sawing followed by herbicide application in the case of coppicing species, or felling, stacking and burning.

Biological control is dependent upon the introduction of natural predatory organisms from the country of origin, but in two cases pathogenic fungi have appeared apparently *de novo*! Introduced organisms are tested very carefully before being released in the veld, to ensure that they will not, in any circumstances, attack related indigenous species. Thus it has been shown that the insects attacking hakeas will die rather than convert to proteas. However, it is understood that research into biological means of acacia control has been restricted to organisms which attack seeds or flowers, after complaints by the Wattle Research Institute, who feared that the *Acacia mearnsii* plantations might be endangered. Incidentally, biological control can never eradicate an alien. The best it can do is to slowly reduce its extent until it reaches an equilibrium, so abolishing the invasive properties.

Invertebrate fauna of the Cape Peninsula mountains

DR VINCENT WHITEHEAD

As a group the invertebrates are highly diverse in form and lifestyle. They are found in most terrestrial and aquatic habitats and have been estimated to make up 95 per cent of the earth's animal species. The simplest forms, the PROTOZOA, are microscopic in size and feed on bacteria, algae and smaller protozoa. Other forms contain chlorophyll and obtain their nutrients by photosynthesis. Parasitic members of this group, which cause several diseases of man and other animals, obtain nutrients from their host by absorption through their body wall. Sponges are multicellular animals which establish colonies by secreting calcareous or silicaceous material held together by protein fibres. Most are marine but there are two species which inhabit fresh water. A closely related group are the jellyfish and corals which are also mainly marine but do have a few fresh-water forms. In contrast to these lower invertebrates, the true worms are distinguished by having a definite head and a bilaterally symmetrical body. They are free-living in marine, fresh-water and terrestrial habitats but many are parasites, such as flukes, tapeworms and nematodes. The segmented worms, the group to which earthworms and leeches belong, are abundant in the streams of Table Mountain and two families of the former are endemic to this range.

In this brief introduction to the lower invertebrates, it is obvious that many are small or cryptic to the extent that they are not readily seen by the casual passer-by. However, there is a large assemblage of higher invertebrate forms with jointed legs, known as the ARTHROPODS, which are generally larger and consequently more obvious. Terrestrial arthropods comprise a large body of animals which include crustaceans, arachnids, myriapoda and insects as well as the enigmatic velvet worms which have affinities with both the arthropods and the true worms. The insects, that is, those arthropods with six legs and three divisions to the body, are extremely abundant and on a global scale make up three-quarters of all living things. Three-quarters of the insects in turn are made up of beetles, which prompted the remark by the famous scientist J B S Haldane when asked by a reporter whether his lifetime study of nature had given him any clues about the nature of the Creator: "He must have had an inordinate fondness for beetles." This generalisation will also apply to the Table Mountain range, but because many beetles are cryptic or live in hidden habitats they may appear to be less common than those insects which live or forage in the open such as bees and butterflies.

The insects are not discussed here in the systematic order found in textbooks but rather according to how frequently they are likely to be seen by the walker or climber. Flying insects such as the butterflies, bees and wasps which are attracted to flowers, are discussed first, followed by the less obvious groups such as beetles, bugs and others.

Butterflies (Lepidoptera)

Because of their attractive colours and habit of taking nectar from a large variety of flowers, butterflies attract immediate attention. On the Table Mountain range 53 species are regularly found, 20 per cent of which belong to the group known as the coppers and blues (LYCAENIDAE). These are small to medium-sized and are less obtrusive than the larger more conspicuous monarchs, swallow-tails, acraeas or browns.

The African monarch (*Danaus chryssipus*) is a conspicuous, slow-flying, predominantly brown butterfly with distinct black and white markings on the wings. Like all butterflies they take nectar from a variety of flowers but females can often be seen hovering over milkweed (*Asclepias*), the plant on which their larvae feed. Eggs are laid on the leaves and the larger larvae, which are mostly white, have yellow and blue-black bands with paired black fleshy projections on their second, fifth and terminal segments. The pupa, which is attached to the plant, is often a striking green with a golden girdle and golden spots on the body. Both adults

African monarch butterfly (*Danaus chryssipus*).

and larvae are reported to be toxic to most vertebrate predators. The leisurely flight of the adult butterfly and the strong contrasting colour patterns on the wings are indications that it is distasteful and not readily attacked by predators. Because of their unpalatability, this butterfly and the garden acraea are copied or mimicked by other palatable butterflies.

The toxicity, which is acquired from their host plant, does not however protect them from insect parasites. Small white cigar-shaped woven cocoons of a small parasitic wasp can sometimes be seen attached to the outside of an obviously ailing caterpillar. These cocoons are spun by the wasp larvae which have fed inside the host and when mature bore out through the body wall and attach themselves to the caterpillar, pupate and become adult wasps. In the process they destroy their host.

Many species of acraea occur through Africa but only the garden acraea (*Acraea horta*) is found on Table Mountain. In this species, as in other members of the genus, the outer half of the front wing is transparent where the scales are absent. The base of the forewings and the whole of the hind wings have light brown to reddish scales with black markings. Male and female are similar except that the colour of the male tends to be more reddish. The flight of this butterfly, like that of the monarch, is slow and indicates its unpalatability, which it acquires from its host plant, the wild peach, *Kiggelaria africana*. In spring batches of yellow spindle-shaped eggs are laid on young leaves at the tip of shoots. The emerging larvae at first remain in a cluster and their feeding destroys the leaf surface, leaving it like a piece of parchment. As the larvae grow they become black with three rows of branched spines and no longer feed together but scatter throughout the tree and may defoliate it entirely. The larvae of the garden acraea have also been shown to be poisonous, taking up cyanide from the host plant; and the only vertebrate known to feed on them is Klaas's cuckoo. Invertebrate parasites are, however, not affected and larvae and pupae are heavily attacked by several species of small wasps, some of which emerge from parasitised larvae and can be seen as small grey oval capsuleas dangling from twigs where larvae have been feeding. Other parasite larvae emerge from caterpillars at the pupation site to spin delicate oval cocoons next to their dead hosts.

The Table Mountain beauty (*Aeropetes tulbaghia*) is the largest of the group of butterflies known as the browns (SATYRIDAE), and has a wingspan of more than 8 cm. They are generally found in mountainous areas and in the vicinity of Table Mountain may come down to lower altitudes to feed on garden flowers. They are

Table Mountain beauty (*Aeropetes tulbaghia*). (*Photo: H Bourn, Cape Nature Conservation*)

particularly fond of red flowers such as red-hot pokers, watsonias and aloes and are reported to be the main pollinator of the red disa. Eggs are scattered among the grasses on which the larvae feed and young larvae may be found here but larger caterpillars shelter among the roots during the day and crawl up to feed at night.

At least nine other species of satyrid butterflies are found on Table Mountain. All feed on grasses and often occur in large numbers, particularly in autumn, flying back and forth over grass meadows. They are dull, light to dark brown, some ornamented with yellow, orange or red dots, and all having eye-spots on the forewings, with some also having them on the hind wings.

The citrus swallow-tail or Christmas butterfly (*Prin-*

Citrus swallow-tail or Christmas butterfly (*Princeps demodocus*). (*Photo: H Bourn, Cape Nature Conservation*)

Yellow dotted border butterfly (*Mylothris chloris*). (*Photo: V B Whitehead*)

ceps demodocus) is another conspicuous butterfly which is common during the summer months. The wings have yellow markings on a black background, with two blue and orange eye-spots on each hind wing. The larvae feed on several plants of the citrus family and also on the unrelated fennel and wild celery. They are highly variable in colour from dark green with a lighter underside to a pale yellow with a brown stripe. When disturbed the larva extrudes two fleshy horns from the base of the head, at the same time emitting a strong citronella odour. The upright pupa, which may be green or fawn, is typical for the family, being attached by the tip of the abdomen to a twig and stabilised by a silken girdle about its middle.

A recent introduction to the Western Cape and Table Mountain is the bright yellow dotted border (*Mylothris chloris*). Previously known only as far south as Knysna it has since 1980 been recorded further and further south, until by 1984 large numbers of larvae were found feeding on Cape sumach (*Colpoon compressum*) at Muizenberg and St James. It is now well established in the Cape and can be seen at most times of the year if the weather is suitable. The wings, which are a rich yellow with orange on the underside towards the base, have a series of black dots along the margins. The pupa, which is attached to a leaf, is pale straw-coloured with projections from the head and sides, and like most butterflies it is also attacked by wasp parasites.

The family with the largest number of species recorded from Table Mountain is the LYCAENIDAE (coppers and blues) with 28 species, and includes the only butterfly species which is restricted to this area. Some are extremely localised and nearly all are associated with ants during part of their larval stage. Some are completely dependent on ants for their survival.

Many of the larvae associated with ants have a large dorsal honey-gland which produces a substance similar to honeydew and which is eagerly sought by ants. At the end of the abdomen some species have an eversible organ which may have some communication function and may possibly be used to deter overexploitation of the honey-gland. Larvae of lycaenids can be grouped according to their relationships with ants. The majority of species in the larval stage feed on plants. Of these some are completely plant-feeding whereas others, after some development, change to a different diet, including the larvae of ants in whose nest they shelter. Another group feeds only on plant-hoppers and may or may not be associated with ants. Others, again, feed on honeydew secreted by plant-lice, and finally there is a group which feeds only on ant brood.

One of the largest lycaenids, the protea scarlet (*Capys alpheus*), in the larval stage feeds in the buds of several protea species. Fire and the resultant reduction of proteas has led to the scarcity of this butterfly on Table Mountain today. A protea host such as *Protea repens*, the common sugar bush, was once extremely common on Signal Hill but no longer grows there; the waboom (*P. nitida*) is, however, still numerous and is now the main host species.

Moths (Lepidoptera)

Willow emperor moth (*Gonimbrasia tyrrhea*). (*Photo: V B Whitehead*)

Larva of the willow emperor moth. (*Photo: V B Whitehead*)

Although there are many more moths than butterflies, most are nocturnal and are seldom seen during the day. Two of the largest moths found on the mountain belong to the emperor moth family (SATURNIIDAE). These are large with a wing span of from 5-10 cm and have a clear window in each wing which is typical for the family. The pine emperor moth (*Nauderelia cytheria*) is predominantly yellow with a straight outer wing band whereas the willow emperor moth (*Gonimbrasia tyrrhea*) is greyer with a wavy outer wing band. The large, spherical cream to white eggs are laid in clusters on the host plant and the larvae when fully grown can be as long as 10 cm. The pine emperor moth has a dark larva covered in bands of blue, yellow and red scales; these differentiate it from the willow emperor moth larva which has only white scales. The larvae of both moths feed on a wide range of host plants which include *Protea*, *Rhus* and *Diospyros*. Pupation takes place in the soil without the formation of a cocoon, and moths emerge in autumn.

Another group of nocturnal moths, the hawk moths (SPHINGIDAE), are attracted to light at night, but are often encountered in the larval form during the day. The lesser arum hawk (*Hippotion eson*) has larvae which feed on arums as well as *Fuchsia* and *Impatiens*, and the large chocolate brown or sometimes green larvae with a down-turned "tail" can be found low down on plants during the day. Large pieces eaten out of leaves or the complete disappearance of young plants accompanied by large pea-sized faecal pellets on the ground is a pointer to their presence.

Bees, wasps and ants (Hymenoptera)

The HYMENOPTERA like the butterflies are some of the more conspicuous insects seen feeding on flowers.

BEES (APOIDEA)

Because they forage from early morning to late in the evening and sometimes even in inclement weather, honey-bees (*Apis mellifera*) are readily seen by the climber, hiker or casual observer. This is the only bee in the southern part of the country that produces and stores honey but further north, in the region of the tropics, the stingless or Mopani bees also store honey and pollen in separate cells for brood provisioning.

In the Southern Cape we have a geographic variation or subspecies known as the Cape bee, which differs from honey-bees in the hinterland in being relatively mild-tempered. The killer bees of newspaper reports and the horror movies are the more northerly variety which, when aroused or interfered with, particularly during hot weather, can cause havoc by stinging poultry and other livestock and have even caused human fatalities. If a hive is found in the veld, avoid it by moving away slowly; and if one is attacked, cover the head and face with clothing and move away quickly. It has been shown that the incidence of stinging decreases rapidly with distance from the hive.

Our southern bees also differ in that the workers can lay fertile eggs. The honey-bee has a highly organised social system with a queen, which lays eggs; drones, which fertilise her on her nuptial flight; and workers that carry out the maintenance of the hive. In the HYMENOPTERA, males result from unfertilised eggs and females from fertilised eggs. Worker bees are females but because they are suppressed by hormonal secretions of the queen their reproductive organs are atrophied. In the absence of a queen this suppression is released and workers are able to lay eggs. Under these circumstances normal honey-bee workers can lay only unfertilised and consequently haploid eggs. These give rise to drones only and in time the swarm will degenerate. This is not the case with the Cape honey-bees, however, as they are able to lay diploid eggs which develop into workers.

While there is only one species of honey-bee in southern Africa, there are many other sorts of bees, probably in the region of 800 species, of which all but a few are solitary.

The allodapine or small carpenter bees (*Ceratitinae*) are partially social bees, and six or seven species have been recorded on Table Mountain. They nest by tunnelling down into the pithy stems of *Aristea*, *Cotyledon*, *Leucadendron*, *Othonna*, *Pelargonium*, *Watsonia* and the stems of other plants with a soft pith. These are small, shiny black to dark brown bees, 4-9 mm long with yellow markings on the head and sometimes also on the thorax. They are active practically all the year round and collect pollen from a variety of flowers; they can often be seen on ericas in the late winter and spring. Adults emerge from midsummer to autumn and live for about a year.

The carpenter bees (XYLOCOPINAE) are large, noisy, black to black and yellow bees, often incorrectly referred to as bumble-bees. The latter, however, do not occur in southern Africa. There are several species on Table Mountain, the large black shiny *Xylocopa capitata* and the slightly smaller, yellow-banded *X. caffra* being the most common. Both nest in dead branches of a variety of trees, and the yellow-banded carpenter bee sometimes uses aloe stems and structural timber. Like the small carpenter bees, they exhibit a partially social behaviour. The adult females are fairly long-lived and after egg-laying has been completed and the brood mature, assist with nest maintenance. The old female cleans and protects the nest and will feed her adult progeny even after they are able to fend for themselves.

Carpenter bee (*Xylocopa capitata*) on aspalathus. (*Photo: V B Whitehead*)

Adult carpenter bees visit many kinds of flowers but the salvia and pea families are usually preferred and it has been found that the large black *X. capitata* synchronises its appearance in summer with the flowering period of the keurboom (*Virgilia*).

Both the large and the small carpenter bees belong to the bee family ANTHOPHORIDAE. Other groups in this family are the very large cosmopolitan genera *Anthophora* and *Amegilla*, which are well represented in South Africa and also on the Peninsula. These are swift-flying bees with round, striped abdomens and can be recognised by the high-pitched whine as they hover in front of a flower. They dart in briefly to take nectar or pollen, dart out to hover in front of the next flower to be visited and repeat the process. They are essentially early morning or late afternoon foragers but can be seen at all times of the day when it is overcast. The flowers most often visited are those of the salvia, potato and tomato families. Burrows, with their built-up tubular opening, are made in compact soil, but where the ground is too sandy they are made in old termite mounds.

Another group, the leafcutter bees (MEGACHILIDAE), are also well represented in the area. Many have large mandibles and all transport pollen on the underside of the abdomen and not on their hind legs as most other bees do. One part of this family cuts out circular pieces of leaf which are rolled and pushed into a previously made tunnel to form the lining. Cells are then made in this tube of dried leaves. These bees often make use of beetle borer holes in wood or nail or screw holes in old structural timber. Nests of the leafcutters are often visited by the beautiful shiny, emerald green, cuckoo wasps which enter the tunnel while the leafcutter is away to lay their eggs in the prepared cells.

WASPS (VESPOIDEA)

How does one differentiate between bees and wasps? In general bees are more rounded and covered with a fairly dense pile and usually they move rapidly from flower to flower. Wasps, on the other hand, are elongate and shiny with a sparser body covering. They move more slowly on the flowers. The only real morphological difference is that bees have branched hairs on the body whereas those of wasps are never branched. Most wasps are predators, feeding mainly on other insects, but there is one group which feeds on pollen. Bees, on the other hand, obtain their protein solely from pollen. Like bees, wasps require nectar for energy and they are, therefore, often found in flowers.

The large black, metallic, narrow-waisted mason wasp (EUMENIDAE) is conspicuous on flowers, particularly umbelliferous flowers, in spring. The oval, domed mud nests are built in sheltered positions on boulders or on window-ledges and under eaves in built-up areas. The multicelled nests are provisioned with paralysed moth caterpillars and after an egg has been laid, each cell is sealed. The resultant larvae feed on the caterpillars and when mature go into a resting stage to pupate and emerge as adults the following spring.

The vespid wasps are related to the eumenids. Adults also roll up their wings when at rest but differ in that they are social and build paper nests. The most common of these are the hornets of the genus *Polistes*, which are reddish-brown with a band on the abdomen. The nest is a single-combed, grey, papery structure hanging from a central stalk attached to a branch, overhanging rock or in built-up areas from the house eaves or verandah ceiling. The nest is started by a single fertilised female, that builds a few cells, lays a single egg in the bottom of each and feeds the growing larvae on masticated insects. When these are fully grown they spin a white cap over the end of the cell to pupate and finally emerge as workers, that help the foundress female as she becomes the queen and is responsible for egg-laying only. The nest is enlarged by the workers, which also feed the growing larvae. In return the larvae secrete a fluid which is eagerly taken by the workers. Larvae which are reluctant to supply this fluid have been reported to receive rough treatment from their nurses by being partially pulled out of their cells and pushed back, after which their oral secretions are usually forthcoming. As the summer progresses, the colony becomes quite large and in autumn slightly larger females and yellow-faced males are produced. They mate and the fertilised females disperse to found their own nests.

Another related species is the aggressive European wasp (*Vespula germanica*), which was accidentally introduced to South Africa and was first seen in the Gardens in Cape Town in 1974. At the moment it is confined to the Cape Peninsula and is particularly abundant in the Kirstenbosch Botanic Gardens. These wasps are a conspicuous yellow and black with a hairy thorax. They take nectar from any flower with nectaries shallow enough for them to reach with their relatively short tongue, and, like the polistes paper wasps, feed their young on masticated insects. They can often be seen flying low over lawns in search of caterpillars.

The nest which is started by an overwintered fertilised female, is situated in a sheltered site such as the inside of hollow trees, eaves of houses, cavity walls and holes in the ground. In an unrestricted cavity the nest is a sphere of several layers of horizontal combs with the cell openings pointed downwards. The whole is enveloped in thin layers of grey wasp paper made by chewing weathered plant material and applying it, moistened with oral secretions, in a series of fan-shaped segments. The European wasp comes from a cold climate where the colonies die out during the winter, but because of the mild climate on the Peninsula they can survive for more than one season.

In general the European wasp is beneficial, as it consumes large numbers of harmful insects, but it captures any suitably sized insect and has been known to kill honey-bees and steal their honey. In autumn when the colony is enlarging rapidly and nectar and insects are becoming difficult to procure, workers are a great nuisance at picnic sites and open-air eating-places. They take jam from scones and meat from sandwiches and can be dangerous if they enter a cool-drink container and are swallowed. A sting in the throat can have unfortunate consequences. The sting, although not as painful as that of the honey-bee, may need treatment in people who are sensitive. Unlike that of the honey-bee the sting is not barbed and can be used repeatedly.

The sphecid wasps (SPHECIDAE) are an extremely large and varied group of sand-nesting wasps. One of these is the slender-bodied black and red *Ammophila*, which has an abnormally long petiole or waist, which is usually a dull reddish-brown in contrast to the rest of the body, which is black. Burrows, which are often dug in compact soil at the side of a path, are about 5 cm deep and end in a hollow chamber. Soil is carried out and kicked away with the hind legs but large lumps of soil or small stones are carried in the mandibles and dropped

Golden sand wasp (*Stizus chrysorrhoeus*) with grasshopper prey. (*Photo: V B Whitehead*)

some distance from the nest. The nest entrance may or may not be closed with a small stone when the female goes out in search of moth larvae. The host larva is stung and paralysed, slung under the body and dragged into the cell at the end of the burrow, where an egg is laid on it. The developing wasp larva is progressively fed by the addition of more parasitised larvae and the tunnel finally closed when the wasp larva is mature.

Spider wasps (POMPILIDAE) are readily recognised by their metallic dark blue colour and long legs, curled antennae and smoky wings. They can often be seen running on the ground, wings flicking, searching for their prey. In the Peninsula the large *Hemipepsis* spider-hunter can sometimes be seen walking backwards dragging a paralysed female of a large hunting spider (*Palystes*). Prior to finding the spider the wasp has sought out some suitable crevice large enough to hold her prey. Where there are buildings nearby the spider is often dragged into the foundations through the air-brick. From time to time the wasp, struggling with her heavy load, will leave the spider and fly back to her destination in the air-brick. Once she has deposited it in the selected crevice, she lays an egg on the spider. It is consumed by the developing grub, which spins a cocoon and remains dormant until the following year, when it emerges as an adult wasp.

ANTS (FORMICIDAE)

This is an extremely large family closely related to the bees and wasps. All ants are social and are distinguished by their narrow segmented waist and elbowed antennae. There are in the region of thirty species on Table Mountain, but only the more common ones will be discussed here.

A colony may consist of less than ten or may have many hundreds of individuals, depending on the species. Most ants build permanent nests in the soil but hollow stalks, logs, and existing crevices in rocks are also used. Others dwell in trees or shrubs in nests built of chewed wood and soil particles cemented with saliva; and some live in leaves spun together with silk from their larvae. Army ants are nomadic and only build temporary nests.

Colonies consist of a queen whose sole function is to lay eggs, and a large number of workers which are all sterile females. They vary greatly in size and form, and some called soldiers, with massive heads, are in fact developed to crack seeds. Males are only present when they are produced for mating flights and after the fertilisation of the future queen, they die. The queen then sheds her wings and finds a suitable site to excavate her nest to start her new colony. The brood consists of eggs, larvae and pupae reared in open chambers. They are blind and legless and are fed initially by the queen but later by the workers.

The larvae, like those of the paper wasps, periodically produce a clear liquid from the mouth and also from the anus, both of which are attractive and are solicited by workers. Many other insects live in close association with ants in their nest and mimic the ant brood by producing attractive glandular secretions to make themselves acceptable. These myrmicophiles, as they are termed, will be discussed later.

Ants generally do not have good eyesight and rely on their chemical senses and ability to detect vibrations. They distinguish each other by their complex odour derived from their nests, and special chemicals called pheromones are secreted which serve to notify the members of a colony of specific conditions. These may be trail pheromones leading to a food source, or alarm pheromones to warn of attack or to elicit assistance.

The Argentine ant (*Iridomyrmex humilis*) is an extremely aggressive ant and was first recorded in Cape Town at the beginning of the century. Introduction is believed to have occurred in bales of forage imported during the Anglo-Boer War and they spread rapidly, particularly along the main rail links from Cape Town to the north. They are relatively inactive during the winter and live in large colonies of many thousands of workers and a number of queens with limited numbers of immatures. As the weather warms in early summer, the winter colony breaks up into smaller units and a queen and a number of workers leave to form a summer colony, which never loses contact with the other colonies. There is apparently no nuptial flight in this species but males fly to seek females from other nests. After fertilisation the queen tears off her wings and remains in the nest, so that the number of queens grows as the nest expands but some will later leave the nest accompanied by workers to found nests of their own.

The Argentine ant is seldom found far from human habitation and is also associated with many of the honeydew-secreting plant-sucking insects such as aphids and mealybugs. By attending these, they reduce the effectiveness of natural enemies and the plant-sucking insects often increase to pest proportions.

The spotted sugar ant (*Componotus maculatus*) is a large, conspicuous species with workers up to 1,8 cm in length. They have a one-segmented petiole and the workers have no sting. As they are nocturnal they are most often seen when their nest, usually under a stone, is disturbed. In colour they vary from light brown to black with various light-coloured marks on the abdomen. The nest entrance is surrounded by a low crater of excavated earth. A related *Componotus* species makes its nest in the dried stems of *Gladiolus*, *Watsonia* and *Aristea* or sometimes in deserted woodborer burrows.

Acantholepis capensis, related to *Camponotus*, is one of the commonest ants on Table Mountain. They are brown in colour and are about 5 mm in length. Colonies, which are never large, have no soldier forms but there may be several queens. Winged males of about the same size occur in autumn. Hemispherical grubs of the guest fly *Microdon* are often found on the underside of stones covering the nest.

Anoplolepis custodiens, the pugnacious ant, is also present on Table Mountain, where they nest in bare open spaces and are extremely active during the day, especially in the hot weather. Workers vary from 5-10 mm, have a dark integument covered with fine pale yellow hairs which often make them appear striped in the sunlight. They form large colonies with the nest opening surrounded by the excavated soil which is spread some distance from the entrance. These ants are mainly carnivorous but also attend aphids and soft scales for honeydew. They have the habit of rushing around in a frenzy on hot sunny days and will readily bite intruders in the vicinity of the nest.

The cocktail ants (*Crematogaster*), so-called as they cock up their heart-shaped abdomens when disturbed, are small and black. They have a two-jointed petiole and a functional sting with a spatulate tip which is used to wipe the poison on the objects causing the disturbance. Most cocktail ants are arboreal, nesting in hollow branches or under bark or in ant-carton nests on shrubs. The carton is spherical, grey to black in colour and made of chewed vegetable fibres cemented with mandibular solutions. The walls are thin and papery with irregular cells like a sponge.

The only species of driver ants (*Dorylus*) in the Peninsula has reddish-brown, completely blind workers which vary in size from 2-8 mm. There is no permanent nest and a temporary nest which is made below the soil surface is moved when the food supply in the vicinity is exhausted. The queen is remarkable in that she is huge, being 5 cm long, with an abdomen enormously expanded by her large ovaries. She can hardly walk and it is assumed that she is carried by the workers when the colony moves. Male driver ants are strong fliers with large eyes and are the large brown furry "sausage flies" found at lights at night.

Ant guests or myrmicophiles

In the section on butterflies mention was made of the interdependence of some butterflies and ants. Several other insects also have a close relationship with ants and are dependent on them to varying degrees. These myrmicophiles can be divided into four categories: true guests which are fed and licked by the ants; those that are tolerated or ignored; the hostile intruders; and, finally, those that are parasitic. One of the commonest guests in ant nests on Table Mountain is a small golden fishmoth or bristletail which can be seen scurrying away when the colony is disturbed. Ants ignore them, but they move with the ants when the latter change nests. They have been found in the nests of the pugnacious ant and the spotted sugar ant, and apparently survive by stealing food when ants feed each other.

When stones covering the nest of the small black *Camponotus* are lifted, a small slug-like insect is sometimes found among the ants clinging to the underside of the stone. When fully grown they are about 5 mm long, dull white and hemispherical with a white network pattern on the sides. At the hind end of the body is a brown protruberance on which two breathing-holes are situated. This is the larva of the guest fly *Microdon*. It is not known on what the fly feeds but it cannot survive without the ants and it has been suggested that it scavenges on the ant excrement. Beetles are also numbered among the ant guests and a small seed-like insect of the family COSSIPHODIDAE lives in the nests of the pugnacious ant. Very little is known about its biology and no immature stage has yet been found.

The larger, more robust ant-nest beetles (PAUSSIDAE) are also tolerated in the nest. They are a shiny reddish-brown with antennae highly modified by reduction of the number of segments and the expansion and enlargement of those remaining into bizarre shapes supposedly as a grip for transportation by the ants. The larvae also live in the nest and feed on the ant brood. Both adults and larvae have glandular areas which secrete a substance imbibed by ants. These beetles are entirely dependent on ants but their hosts do not benefit in any way.

Another beetle, *Genuchus hottentotus* of the dung-beetle family, is often found in the nests of cocktail ants. They are attacked by the ants but are protected by their hard, shiny body covering. Like most of the beetles in this family, the larvae feed on material high in organic content and probably scavenge in the ant refuse middens.

Beetles (Coleoptera)

Beetles are the largest order of the insects and are extremely diverse in size and habitat. The group is characterised by having the first pair of wings hard and

Ten-spotted ground beetle (*Anthia decemguttata*). (*Photo: H Robertson*)

Protea beetle (*Trichostetha fascicularis*).

leathery as a protective covering to the delicate flying hind pair.

The staghorn beetle family (LUCANIDAE) is not well represented in South Africa, but *Colophon* of this family is one of the unique insects of Table Mountain. The mandibles of *Colophon* are relatively small and in no way compare with the development of the large antler-like jaws of their European relatives but they do have characteristic shapes which are very useful for identification. What makes this group interesting is that they are only found on specific mountain tops. *C. westwoodi* is found on Table Mountain only, with a different species on the Hottentots Holland, another in the Worcester area and again other species on the Langeberg range. Very little is known of their biology, but from the knowledge of the stag-beetles, which feed in decaying wood, it is thought that they may feed in the bases of *Restio* clumps. No larvae have yet been found and only a few adult beetles have been seen alive.

Less dramatic, but still spectacular, is the flightless, fast-running, 10-spotted *Anthia decemguttata* ground beetle (CARABIDAE). They are large, from 2-3 cm, mainly black, sometimes with a reddish thorax and all with obvious longitudinally ridged wing covers which may be completely black with no spots but can have up to the full complement of ten spots. Many of the species in this group are noted for their defence mechanisms and if threatened can squirt formic acid a distance of up to 35 cm in any direction, hence the Afrikaans colloquial name "oogpister".

The scarab or dung-beetle family (SCARABAEIDAE) is extremely large and is made up of such well-known groups as the dung-beetles, fruit-beetles, rose-chafers and monkey-beetles. The green protea beetle (*Trichostetha fascicularis*) is well known for its visits to large protea flowers such as *Protea cynaroides*. This green-backed beetle with its bright yellow clusters of hairs along the sides is quite common, but there is also another colour form which has chocolate-coloured wing covers and pale hair bundles along the sides. Larvae of this whole group feed on rotting plant material and the protea beetle has been recorded from thatched roofs at Constantia where the thatch had begun to rot and also from nests of the black mound termite where it fed in their refuse middens.

Related to the protea beetle is the sinuate fruit-beetle (*Panchnoda sinuata*). This compact, shiny yellow and black beetle has recently arrived in the Peninsula from the Karoo. Collection records show that over the last 25 years it has gradually spread from Worcester throughout the Boland and finally reached the Peninsula in 1979. Adults feed in flowers or in soft fruit and are regarded by gardeners as a pest, while the larvae feed mainly in manure or compost heaps. The white C-shaped larvae can be recognised by the habit of scurrying away on their backs when disturbed. Increased making of compost on farms and in suburban gardens probably helped to cause this recent invasion.

Some species of monkey-beetles (HOPLIINAE) are brown to black and covered in long hairs; others have a short pile often with coloured scales; all have extremely long hind legs with large claws. During spring at the peak of the flower season they can be seen flying or clustered in practically any flower, yellow flowers of the daisy family being particularly attractive. Females burrow into the soil to lay eggs and the larvae feed on organic material or possibly plant roots.

Rose-chafers (RUTELINAE) are related to the previous group but fly mainly in the evening or at night. They are dark brown to bronze-coloured and often come to lights in warm summer evenings. Adults feed on foliage and leave a skeleton of veins. Larvae feed on roots and are known to cause extensive damage.

The long-horned beetles (CERAMBYCIDAE) are medium to extremely large beetles with long antennae. The immature stages are all borers in wood, some being highly specific. *Promeces* is medium-sized and because of its brilliant metallic blue-green colour is often noticed flying in early summer in characteristic slow manner with antennae out and wing covers lifted. They settle to feed in umbelliferous flowers but can also be found on recently felled partially dry timber, where they lay their eggs. The large black and red *Ceroplesis* is a common species in the Peninsula and attacks a wide variety of woods. Oak wood stored for winter use often yields a crop of this species. The most notorious of the long-horned beetles in the Cape is the Italian beetle, which was accidentally introduced from Europe. The larva feeds in pine timber only; it is the reason for the regulations regarding the inspection of roof timbers of older houses. All structural timber of pine is now treated against this insect. The adult is rather nondescript, 1,5 cm long, dull brown with two small light-coloured marks on the wing covers.

The MELOIDAE or blister beetles are also known as CMR beetles, as their colouring of yellow and black is that of the old Cape Mounted Rifles. In the adult stage they are voracious feeders of flowers. All have extremely interesting life histories, as the immature stages are predatory on eggs of grasshoppers, solitary bees and ground-nesting wasps. Eggs are laid in the soil and the first-stage larvae, which have three-clawed feet, are highly mobile, climb onto flowers and attach themselves to a visiting bee, and are transported to the nest. The larva enters the nest and feeds on the contents. Those that attack locust egg-pods search out and burrow into the egg masses, become sedentary and when mature go into a resting stage. They may remain dormant for up to 4 years before pupating and emerging as adults. Most adults produce a poisonous substance, cantharidin, which is highly irritating and in some species may cause painful blistering.

The ladybird beetles (COCCINELLIDAE) can be divided into three groups: those that feed on plants; the fungus-feeders; and the largest group, which are predatory. On the Peninsula one of these plant-feeders (*Epilachna undulata*), with light wavy markings on the wing covers, feeds on *Arctotheca* and *Tetragonia* on beach dunes. A larger species, pale brown with dark spots, feeds on cucurbits. *Leonotis* and members of the potato family are attacked by another of the plant-feeding ladybird beetles. Larvae of this group can be recognised by their covering of stout branched setae.

Psyllobora variegata is a small lemon-yellow and black fungus-feeder which becomes quite numerous during autumn on plants harbouring honeydew-producing insects as the sooty fungus which grows on the honeydew provides them with a plentiful supply of food. Of the predatory species the most often noticed is the large red and black lunate ladybird *Cheilomenes lunata*. This species, as well as two yellow and black spotted ladybirds, feed on aphids during the adult and larval stages. One of these yellow and black species, *Hippodamia variegata*, is extremely variable with regard to the spot pattern on the wing covers. It arrived accidentally in the Western Cape in the late 1960s and within 2 years had spread through the whole of southern Africa. During the dry part of the year they tend to congregate in fairly large numbers in sheltered sites and remain dormant until cool weather arrives, when aphids again become available. This species can be readily recognised as it is much longer than it is broad, while most indigenous species are almost circular.

Lunate ladybird beetle (*Cheilomenes lunata*).

Although there are other organisms, both plant and animal, which give off light, the fireflies (LAMPYRIDAE) are one of the best known. The light they produce is generated in a special organ, usually visible at the tip of the abdomen and appearing as a pale yellow area in dead specimens. The exact position varies somewhat and depends on sex and species. The flashing, which occurs in some species, is brought about by the pulsed regulation of the oxygen supply to the luminescent area.

There is an enormous variation between species in flashing sequence, colour of light, time interval and duration of flash, but there are basically two systems. In one group, where the female is larviform and sedentary and the male is winged, a species-specific signal is broadcast by the female and is responded to by the highly mobile male. In the second system, where both sexes are winged, one sex, usually the male, broadcasts a species-specific signal and the other sex responds, usually from a sedentary position. On the Peninsula we have members of both groups: the LAMPYRINAE, with flightless females which glow (glow-worms), and winged males; and the LUCIOLINAE, in which both sexes flash, the males with a luminous organ of two segments and the female with one. In the glow-worm group females are wingless and black, sometimes with light markings along the sides. They have the appearance of a segmented slug but have three pairs of legs. In the other group both sexes are fully winged. Most fireflies feed on slugs or snails but some adults may not feed at all. Those which are winged and are luminous in both sexes, are known to be active in the Kirstenbosch Botanic Gardens during warm early summer evenings, and males of a glow-worm were attracted to a solitary candle in a darkened room at Hout Bay.

Flies (Diptera)

This order is easily recognised by having one pair of transparent wings and a pair of opaque balancers in place of the second pair. In warm weather in urban areas the housefly and some related (muscoid) flies can become plentiful. When one gets away from built-up areas these flies are less abundant but can occasionally be seen taking nectar from umbelliferous flowers; and if a decomposing animal is found the metallic blue/green blowflies and the large flesh-flies with striped thorax will be found at the carcass. They deposit eggs which in a very short time hatch into cream-coloured maggots that reduce the carcass to a bundle of fur and bones in a matter of days.

We do not have the same fly problem as Australia. Indigenous dung-beetles of Australia can handle the coarse pellet-like dung of kangaroos and other marsupials but cannot consume or bury the large amounts of dung of introduced cattle. In southern Africa we have a variety of dung-beetles which have evolved with our game and are capable of breaking up bovine dung rapidly, which limits the breeding of flies. It was only after the introduction of dung-beetles to Australia from Africa and India that some relief of the fly problem was obtained.

The robber flies (ASILIDAE) have a noisy buzz and only fly a short distance before settling again. There are many species varying in size from extremely small to greater than 2,5 cm. Adults have a bristly face, pointed mouth parts and a long and sharply tapering abdomen. They tend to be territorial, and dart after prey in the vicinity, pierce them with the sharp mouth parts and land nearby to feed.

During early spring and summer several blood-sucking flies (TABANIDAE) can be a nuisance in the field and their bites result in a large itchy swelling the following day. Near streams or moist ground one is often attacked by "blind" flies. They are dull grey and settle on one silently, the sharp pain of a bite being the first indication of their presence. They are easy to swat, as they are slow-moving and poor fliers. More agile but with a bite just as painful are the small tabanids with metallic-coloured eyes. They are active when the spring flowers are out.

Another group of annoying flies are the biting midges or "muggies" (CERATOPOGONIDAE). They are small (3-5 mm) and are most numerous in the spring, usually near water or damp areas. At their peak they can form a comet-like swarm around one's head, getting into the eyes, ears and hair. Only females bite and some species inject an anaesthetising substance when they bite and the effect of their attack is only felt the following day when the affected areas swell and itch.

False mosquitoes (CHIRONOMIDAE) are aquatic flies that often emerge simultaneously on the first warm days of spring creating large and often alarming swarms. In the evenings they are conspicuous against the setting sun and as they are attracted to light, they can be a nuisance in the home at night. In some areas of California they occur in such large numbers that they have been known to put out camp-fires at lake sides. Their resemblance to mosquitoes can be alarming but as they do not feed on blood, they are harmless.

Crane-flies or daddy-long-legs flies (TIPULIDAE) abound in damp situations and in appearance remind one of large long-legged mosquitoes but adults are harmless as they do not feed. During the day they rest in the plant cover and in the evening can be seen bobbing up and down close to the soil surface as they lay eggs.

Bee-flies (BOMBYLIIDAE) are frequent visitors to flowers. These hairy bee-like flies make a persistent high-pitched whine when hovering and can often be seen standing on the flower and probing for nectar. The larvae are predaceous and one species is known to feed on carpenter bee larvae.

Hoverflies (SYRPHIDAE) are medium to large flies often with colourful banding on the abdomen. They have the characteristic habit of hovering in the vicinity of flowers or plants infested with aphids. Syrphid larvae are slug-like, often green or brown in colour, with tubercles on the body. They feed on the aphids by piercing the soft skin and sucking out the body fluids. If such a larva is closely observed it will be noticed that the area around it has been cleared of aphids by its feeding.

The members of another group of this family are known as drone flies or rat-tail flies. They are found in wet or marshy areas where larvae feed in stagnant eutrophic pools. Because of the lack of oxygen in such habitats, the larvae have developed a long tubular extension on the hind portion of the body which reaches to the surface for air. Adults are bee-like and can often be seen taking nectar from flowers with shallow nectaries.

Crickets, grasshoppers and locusts (Orthoptera)

This order consists of groups: the crickets and their relatives with long antennae (more than thirty segments); and the grasshoppers and locusts with short antennae.

The crickets include such insects as the common garden cricket, katydid, mole cricket, ghost cricket and many others. Most are nocturnal and all stridulate. The forewings act as a shield for the delicate underwing and are the main stridulating organs, being lifted up and rubbed against one another.

On summer evenings a low continuous buzzing from the lawn or some damp area indicates that a mole cricket (*Gryllotalpa africana*) is stridulating at the entrance of its burrow. Mole crickets, as the name implies, burrow, using their powerfully clawed front legs. They are a light brown with the underwings rolled up and projecting like two spines past the end of the abdomen. A high-pitched continuous or sometimes interrupted song usually emanates from the pale green ghost crickets (*Oecanthus*), of which there are two in our area. They are difficult to locate when stridulating as their sound is unidirectional and when reflected off a leaf gives the wrong direction to the whereabouts of the cricket. Some species of this group use natural openings between two adjacent leaves through which to stridulate and in so doing use the principle of a sounding-board to amplify their sound several times. Others in this group are the katydids, which are usually green, have extremely long antennae and start stridulating in the early evening. Their song is in the form of a short chirp which has the sound of "katydid". The cave cricket of Table Mountain (*Speleacris tabulae*), like the mountain-top beetle (*Colophon westwoodi*), is only found on the Table Mountain range. They are dark and long-legged with extremely long feelers and are found in the innermost corridors of caves where little light reaches. They have been found feeding on a dead bat and probably also feed on bat dung and lichens.

Locusts and grasshoppers have short antennae of less

Cape ghost cricket (*Oecanthus capensis*). (*Photo: V B Whitehead*)

Foam locust (*Dictyophorus spumans*).

Bladder grasshopper (*Bullacris discolor*). (*Photo: Moira van Staaden*)

than thirty segments. Wings are usually well developed but short-winged and wingless forms do occur. The term locusts is usually reserved for those species which under favourable conditions congregate in large swarms where they change colour and develop longer wings than normal. Despite this distinction many of the large non-swarming members of this group are also called locusts. The rather large gaudy green and red bush locust

(*Phymateus leprosus*) and the large green, yellow and black milkweed locust (*P. morbillosus*), although numerous at times, never swarm but they do congregate in the hopper stages. Both these species are slow-moving and feed in exposed sites on top of shrubs. This behaviour and their strongly contrasting colours are a warning to would-be predators that they are poisonous.

The bladder grasshoppers (PNEUMORIDAE) are a peculiar group of short-horned grasshoppers restricted to southern and eastern Africa. They are often cryptically coloured, being green or brown, sometimes with white or silver markings. *Bullacris discolor*, which is quite common on the Peninsula, is a uniform pale green but brown or light pink specimens also occur. Males, which have an enormously expanded abdomen, are winged and are quite strong fliers. They stridulate by rubbing the hind legs against a row of teeth on each side of the abdomen and produce a loud sound which can be heard for several kilometers. Females, which are wingless and slow-moving, apparently also stridulate but to a lesser degree.

Mantids (MANTODEA) are commonly called praying mantids, from the way the front legs are held. In the mythology and beliefs of the Khoikhoi the mantid was seen as an embodiment of one of their deities, and as such was a bearer of good omen and an oracle to be consulted. Mantids blend with the background and, depending on the species, may be green or light brown to dark brown. A large green species is often attracted to insects fluttering around lights at night. Two characteristics, the raptorial front legs which close on the prey like the blade of a pocket knife, and the long narrow thorax, distinguish these insects. Not all are winged but in those that are, the wing covers are long and narrow and the underwing is folded and membranous. Eggs are deposited in a frothy mixture which hardens into a tough straw-coloured egg packet characteristic in shape for each species. Young mantids emerge from the top of the egg packet, but holes along the side are those left by a wasp which parasitises the eggs.

Female praying mantid feeding on a fly. (*Photo: V B Whitehead*)

Termites (Isoptera)

The "anthills" so characteristic of much of the African landscape, are the aerial part of termitaria, the homes of termites or "white ants". They are very ancient insects, having survived almost unchanged for over 100 million years, the termite mound being the oldest example of an organised community found on earth. More closely related to cockroaches than ants, they are highly social insects with four castes: the dark-coloured winged reproductives; supplementary reproductives; soldiers, which are sterile males and females; and, lastly, the workers, which are also sterile males and females.

New colonies are formed when large numbers of winged males and females are released from the mound. They fly weakly, fall to the ground and break off their wings. The female then lifts her abdomen skywards and emits a scent which attracts a male. They set off in tandem to find a suitable spot to start a nest. They mate and tend the first eggs and early stages of the resultant workers; thereafter the workers take over this duty and the female becomes an egg-laying unit. Plant material is the only food of these insects and they harbour special micro-organisms in their gut to enable them to digest cellulose, and thus play an important role in recycling plant material. Partially digested oral and anal food is passed on to other workers and in this way the food is fully used. The non-digestible black food remains are incorporated into the nest walls.

In the Peninsula the black mound termite (*Amitermes hastatus*) is common and after rains in early spring the mounds are enlarged while the surface is soft. Patches of black, crusty material on the surface indicate that thousands of workers just below the surface are actively extending the mound. Winged individuals can be found in the nest from mid-January but will only emerge as fully winged forms in April or May after the first autumn rains.

The other termite found in this area is the harvester termite (*Microhodotermes viator*), which is not a mound-maker, and nests are indicated by small scattered heaps of dark soil. More obvious are the harvesting-holes where the workers deposit pieces of dry plant material which is eventually taken down into the nest. This species is active during the day and workers are well pigmented.

The true bugs (Hemiptera)

The Hemiptera form a very large order that is divided into bugs which have both wings delicate and transparent (Homoptera) and those whose forewings have the basal part (closest to the point of attachment) opaque and thickened (Heteroptera). All have tubular mouth parts and most live by sucking sap from plants. A few are predatory and feed on other insects or vertebrates by sucking their body fluids.

HETEROPTERA

The cotton-stainers (PHYRRHOCORIDAE), which are bright orange or red, are often seen coupled end to end in copula. They feed on seeds and their preference for plants of the cotton family and the stain resulting from their feeding in cultivated cotton gives them their name. Locally they can be found feeding on malvaceous plants such as mallow.

The large dark twig-wilters which have extended angles to the thorax also belong here. They feed by piercing young shoots or the stem at the base of buds to feed, causing them to wither and die. Like many of the bugs they have several repugnatory glands along the side of the body through which they eject an irritant and foul-smelling fluid as a defence mechanism.

The shield bugs (PENTATOMIDAE), which have a shield-shaped body, often brightly coloured, with well-developed stink glands, are also plant-feeders. They can be found feeding in seed-heads of grasses on coastal dunes or in the clusters of *Rhus* seeds.

Some of the Hemiptera have taken to an aquatic habitat and the most elegant of these are the water-striders or pond-skaters (GERRIDAE) with their narrow bodies and long legs. They are predaceous and feed on any insect trapped in the water surface.

The water scorpions (NEPIDAE) are large (2-4 cm), elongate aquatic bugs with two long abdominal extensions which form the respiratory tubes. They are slow-moving and are found at the bottom of shallow ponds with the breathing-tube at the water surface. The raptorial front legs are used for grasping their prey.

HOMOPTERA

This group includes insects such as cicadas, scales, mealybugs, aphids, leafhoppers and others. Under natural conditions they are not often found in large numbers, indicating that efficient natural controls exist.

The most obvious of this group, audibly, if not visibly, are the cicadas (CICADIDAE). They are cryptically patterned and can be seen only by diligent scrutiny of the area from which their sound emanates. When disturbed they make a short flight and settle nearby. The sound, made only by males, is produced by the rapid movement of a circular membrane on the side of the abdomen. Females lay their eggs in slits made in twigs and the nymphs drop to the ground and burrow into the soil. They feed on roots and take several years to complete their larval life. The final instar emerges from the soil and attaches itself to a tree-trunk before splitting open for the adult to appear. These empty larval skins can be seen clinging to tree-trunks in early summer.

At the same time of the year some grasses appear to have patches of froth on their stems. These are caused by froghopper or spittlebug (CERCOPIDAE) nymphs which enclose themselves in anal secretions frothed up by blowing air through modifications of the abdominal wall. This device may protect the delicate insect from desiccation or even from natural enemies.

In late summer many of the olive trees, usually the shrubby growth at the base of the trunk or young seedlings, appear to be adorned with wisps of bluish-white threads. A closer examination will show that this is a filamentous waxy covering of a small insect feeding on the young twigs. These are the immature states of the olive psylla (*Euphyllura longiciliata*).

Another group of the true bugs are the coccids or scale insects (COCCOIDEA). They are usually small, often flat and inconspicuous. Females are wingless and usually secrete a protective waxy cover over the body. Many of the scales secrete a sweet honeydew which is a by-product from the processing of a large volume of plant sap in order to obtain sufficient protein. This product is attractive to ants and is also a growth medium for sooty moulds.

Included in this group are the wax scales (*Gascardia*) often found on *Dodonaea*; lac insects, which appear as shiny brown encrustations; and the soft scales, which are flat and oval. A group of scales which do not produce honeydew are the cryptic diaspine or armoured scales (DIASPIDIDAE). The circular, dull brown, freckle-like scales sometimes found on oranges and on some garden plants such as roses, and similar scales on aloe leaves, belong to this group. Under the thin wax cover lies the soft female with her fine thread-like mouth parts embedded in the plant tissue. As with the soft scales, this group is also well controlled by its natural enemies and only occurs in large numbers where some disturbing factor, such as road dust, deters their minute wasp parasites.

Aquatic arthropods

On Table Mountain aquatic animals are adapted to the acid peaty brown water. The water is particularly pure and low in nutrients, and the forest trees which overhang it supply most of the organic content. This material would merely be carried away but for the numerous organisms which, by the shredding action of their feeding, break it down and make it available to others.

One of these is the amphipod genus *Paramelita*, which has at least four species endemic to Table Mountain and the Constantiaberg. The river crab, *Potomonautes*, is an African endemic and recent research indicates that there are several isolated groups of the river crab on Table Mountain which can be differentiated from the above genus.

Many insects and other arthropods are associated with permanent running water and play an important part in the ecology of such streams, not only breaking down organic material but in turn providing food for other aquatic organisms. The dragonflies and the related damselflies (ODONATA) are usually found near water, criss-crossing the area to catch prey on the wing. Damselflies are more slender and are able to fold their wings back over the body, in contrast to dragonflies, which always rest with their wings extended. Larvae of both are carnivorous and completely aquatic and grasp their prey with their toothed and prehensile lower jaw.

Less obvious are the stoneflies (PLECOPTERA), which are small cockroach-like insects. Adults are elongate and soft-bodied, with two pairs of similar-sized wings folded back over the body. They scuttle rapidly over rocks and only occasionally fly. In overcast or drizzly weather adults can be found at the sides of streams but during hot weather they seek out cool, shady spots. The larvae are aqautic.

Adults of caddis-flies (TRICHOPTERA) are small moth-like insects with hairy wings. They may be difficult to find as most are crepuscular or nocturnal. The larvae are aquatic and are important to the ecology of streams, both in breaking down organic material and as food for many fish species.

Mayflies (EPHEMEROPTERA) are small, short-lived insects with long caudal filaments; the front wings are large and triangular and the hind wings small and reduced. The gut, which is filled with air to aid buoyancy, is not functional, as adults do not feed. The larvae are aquatic and feed underwater on plants. Most adults live for a few days only and in some species emergence is synchronised, leading to spectacular swarming flights.

Cave-dwellers

The caves of Table Mountain form extensive underground systems which support a unique animal life. Several new species of spiders and millipedes, as well as the cave cricket (*Speleacris tabulae*) whose nearest relative is found in New Zealand, have been described. There is also a cave woodlouse, (*Trichoniscus tabulae*) and an unpigmented cave pseudoscorpion (*Chtoniella cavernicula*). The world's only cave velvet worm, the white *Peripatopsis alba*, is also found here, as well as the crustacean *Spelaeogriphus lepidops*, which is so different that it has been placed in a new order.

Onychophora

Velvet worms, as they are commonly called, are found sparsely distributed in tropical and subtropical areas. In South Africa they occur mainly in the discontinuous forests from the Cape Peninsula to the Eastern Transvaal. Four species, all in the genus *Peripatopsis*, are found on Table Mountain, with two being endemic. They are curious velvety, worm-like creatures with 14

Dragonfly. (*Photo: Zelda Wahl*)

Velvet worm (*Peripatopsis capensis*) giving birth to young. (*Photo: Zelda Wahl*)

to 25 pairs of unsegmented legs each ending in a pair of claws. The body and the antennae are unsegmented. These animals are the remnants of an ancient lineage dating back 500 million years and have characteristics which link them to the true worm and the insects. Although they are not common they are occasionally found in rotting logs or emerge from their hiding-places in deep litter during cool, rainy weather or at night.

Spiders and scorpions (Arachnida)

Other terrestrial non-insect arthropod groups which can be encountered on Table Mountain are the spiders and scorpions. Although many of the spiders are small and cryptic, others are large and often encountered or spin their webs in open situations. The largest are the baboon spiders, which are hairy, black or brown with red around the mouth. When alarmed they rear up exposing this area and showing the large fangs. This rather terrifying-looking animal can inflict a painful bite and although one species is considered to have venom toxic to man, the majority are not considered to be dangerous. They live in silk-lined burrows which may be up to 30 cm deep. The female rarely if ever leaves the nest, waiting for prey to become entangled in the web which she spins over the burrow entrance. Those spiders encountered in the open are usually males which are attracted to the female in the mating season. They take 3-5 years to mature and may live for up to 25 years.

Male baboon spider (*Harpactira baviana*) in threatening pose. (*Photo: Norman Larsen*)

Rain spider (*Palystes castaneus*) guarding her egg sac. (*Photo: Norman Larsen*)

Another large spider is the hunting spider (*Palystes*), also known as the rain spider or wall spider, as it is often found in houses where its appearance at night is supposed to indicate the approach of rainy weather. Females are large and hairy, up to 9 cm across with black banding on the underside of the legs, and although they look alarming, they are not considered to be dangerous. They are free-ranging and no web is spun, but silk is used to construct the nest. This is a large structure about the size of a tennis ball, consisting of leaves loosely held together with silk and suspended from a shrub or tree. Females watch over the nest until the young hatch, after which it is abandoned.

Wolf spiders (LYCOSIDAE) are small to medium-sized spiders which are free-ranging and run down their prey, hence their name. They can be found in many habitats: under stones, in old tree-trunks and in the grass. They can be recognised by the habit of the female of carrying her egg sac attached to her spinnerets and by the three rows of eyes, two eyes in each of the upper rows and four in the lower row. The young are carried on the back of the female and leave after their second moult. The bite of these spiders is considered to be harmless, although there may be some pain and redness at the location of the bite.

The garden orb web spider (*Argiope*) is large and spectacular with its silver and yellow body and black markings. The female with her ribbed and lobed abdomen can be found in the centre of her huge orb web.

One can often see what appears to be a dead honeybee attached to a flower, but closer inspection will show that it has been caught by a small white, yellow or pink spider. This is the crab spider (THOMISIDAE), so called because of its crablike movements. They are small, seldom being larger than 1,5 cm, and are most likely to be found waiting in a flower with the front legs outstretched to seize any visitor to the flower.

Button spiders (THERIIDAE) have acquired a certain notoriety because of the toxicity of their venom. The black button spider occurs throughout South Africa and is usually only found in the veld. The local species, *Latrodectus indistinctus*, can be recognised by the central red stripe down the hind part of the abdomen. This

Crab spider (*Thomisus* sp) feeding on a honeybee. (*Photo: Norman Larsen*)

fades as the spider ages and some mature females may have no colour on the abdomen except a dull red dot above her spinnerets. There is no red mark on the underside of the abdomen. Females vary in colour from brown to black and have a spherical velvety body. Some human fatalities have resulted from the bite of this spider, and children under the age of five or adults suffering from lung or heart complaints are at risk. An antivenom is available.

The related brown button spider or house button spider (*L. geometricus*) is common on the Cape Peninsula. Its colour varies from brown to black but it can readily be distinguished from the black button spider by the brick-red hourglass-shaped mark on the underside of the abdomen. In addition they have a spiked covering to their egg sacs in contrast to the smooth egg sacs of the black button spider. The bite of the brown button spider is not considered to be dangerous but a doctor should be consulted if a young child is bitten.

Four species of scorpions are found on Table Mountain, the largest and most common being the Cape burrowing scorpion (*Opisthophthalmus capensis*), which is found under stones mainly on the western slopes, is large (9 cm), yellowish-grey to brown, with large pincers and a thin tail somewhat longer than the body. The sting causes a painful burning which may persist for several hours and although there may be some swelling the victim will be in no danger. Two other species, both in the genus *Uroplectes*, occur throughout the whole area, also under stones, but sometimes under bark. They are 4,5 cm long, yellowish-grey with thin, smooth pincers. Although they belong to the poisonous buthid family they are not considered to be dangerous and treatment should be aimed at alleviating the pain caused by the sting.

The Cape thick-tailed scorpion (*Parabuthus capensis*) is large, 4-4,5 cm, yellowish-grey to brown, sometimes dark olive grey, with a finely granulated integument and smooth, thin pincers. The tail is thick and longer than the body. In contrast to the three species above, the sting of this species is dangerous and has been responsible for some deaths. Victims should be kept under medical observation and should be treated with antivenom when effects of the poison, such as difficulty in breathing, a drop in the pulse rate and increased salivation, become evident.

It must be appreciated that all the arthropods of the Table Mountain range cannot be discussed, firstly because of limited space but also because of our lack of knowledge. We have found that in any group of invertebrate animals investigated in depth, 50-60 per cent are

Cape burrowing scorpion (*Opisthophthalmus capensis*). (*Photo: Zelda Wahl*)

found to be new to science, which means that on average about half of our invertebrate fauna still have to be described. When we start to look at their biology, that is, how they live or what they do, we find that we have very little information. This is well illustrated by an analysis of a recently published catalogue of all the flies of Africa south of the Sahara. Of the more than 16 000 fly species listed, only 5 per cent of the entries have information on their biology.

As an aid to those interested in learning more about the animals they see, a list of suggested reading is appended. Identification and naming is a highly specialised skill and can only be undertaken by trained scientists who have access to the technical literature, often only available in the libraries of universities, museums or other research institutions.

Vertebrate fauna of the Cape Peninsula mountains

DR DOUGLAS HEY

This chapter endeavours to provide a popular introduction to some of the vertebrates that might be seen on the mountain. It has been written for the layman in the hope that it will add to the enjoyment of walks, stimulate interest in the wildlife of the Cape Peninsula, and so engender public support for the conservation of this unique nature area. Only the commoner species have been described and in the localities where they are the most likely to be seen. Readers requiring more detailed information are advised to consult the relevant reference works listed on page 140.

Mammals

Visitors to the Cape Peninsula must not expect to see the herds of wild animals which have come to be regarded as a characteristic feature of southern Africa. The Cape is characterised by scenic beauty, a pleasant climate and a remarkable variety of wild flowers rather than by its fauna. The Cape flora, despite its rich diversity of species, does not support large populations of animals for any length of time due to its low food value, the absence of suitable grasses and certain trace element deficiencies. In addition the area has been inhabited by man for over 300 years, resulting in the extermination of the larger fauna and the deterioration of their habitat.

The early explorers recorded that elephant, black rhino, Cape buffalo, eland, hartebees and zebra grazed at times on the slopes of Table Mountain; hippopotami wallowed in the vleis of the Peninsula, while lion and leopards raided the livestock of the Dutch East India Company and hyaenas ("wolwe") scavenged around the settlement. As the Peninsula was settled, the fauna was shot out or driven northwards. The last lion was killed on Table Mountain in 1802, eland disappeared in 1840, while leopard, being more secretive, survived until 1870. Of the fauna that were once found at the Cape, 11 species have been exterminated and at least 25 species may be regarded as rare. It is recorded that the ornithologist Le Vaillant shot a Cape vulture on Table Mountain in 1872. The only surviving breeding colony of this once common species in the southern Cape today is to be found on the Potberg in the De Hoop Nature Reserve near Bredasdorp.

The largest and most obtrusive of the approximately fifty species of mammals still surviving on the mountains of the Peninsula today is the chacma baboon; the word chacma being derived from the Khoi name for these animals. They wander freely over the mountains, living in troops of twelve to over fifty animals with a well-defined social order, led by a dominant male. Sentries, usually the younger males, are posted on the outskirts of the troop, whether resting or feeding, to warn of approaching danger. The danger call is a loud, deep, guttural "bog-gem, bog-gem", taken up by other sentries and echoed by the surrounding hills, thus galvanising the troop into full flight with the males bringing up the rear. Both males and females are fearless in defence

Brown hyena ("strandwolf"). (*Photo: Zelda Wahl*)

of the young. Male baboons have exceptionally large canine teeth and are more than a match for any dog.

A baboon troop may range over a wide area, feeding on roots, bulbs, flowers, fruits, seeds, invertebrates and young birds. They have been known to kill an occasional small antelope. In the Cape Peninsula they also scavenge along the beaches and intertidal zone and frequent parking and picnic areas, raiding rubbish bins in search of food. Visitors are warned not to feed baboons as they can become aggressive and dangerous when frustrated. Mountaineers have been attacked while trying to recover their rucksacks, which the animals have learned usually contain food. The troop always returns in the evening to the shelter of high trees or some inaccessible krans, secure from attack by their major enemy, the Cape leopard. With the extermination of leopards, baboons are increasing in numbers and control measures are necessary from time to time.

Leopards still survive in the rugged mountains of the Western Cape. The leopard is a magnificent catlike animal varying in weight from 35-70 kg, marked with black spots and rosettes on a background varying from straw-coloured to deep orange. A striking feature is that even the most brightly coloured leopard blends with the environment when viewed from a distance. It is a solitary animal except during the period when the female is nursing cubs, which remain with her until able to fend for themselves. Leopards hunt a variety of prey including baboons, antelope, field rodents, birds, fish and even crabs. They will also feed on livestock when the opportunity offers. Another noteworthy habit is that of storing their prey high in the fork of a tree – a feat requiring remarkable strength. Once widely distributed over the African continent, Asia and India to Malaysia, leopards have been exterminated from many regions and are classed today as rare.

Cape leopard.
(Photo: H Bourn, Cape Nature Conservation)

A mammal commonly associated with rocky areas of the mountain chain is the rock rabbit – hyrax or dassie. The word rabbit is a misnomer, for the hyrax is not a rodent but belongs to a family of its own with some affinity to the elephants. They are usually to be seen sunning themselves in groups on the highest kranses or along boulder-strewn shorelines. They are agile creatures, the soles of the feet being smooth and well padded and adapted for rock-climbing.

Dassies are gregarious and live in colonies. The males seek vantage points from where they give a harsh warning call at the first sign of danger, causing members of the colony to scurry for shelter. Their senses of sight, hearing and smell are remarkably acute. Dassies feed on a variety of plants but never graze far from their rock shelters. A characteristic of a colony of dassies is a strong aromatic smell from the urine which is usually excreted on the same spot and evaporates to a hard, dark conglomerate of glass-like appearance. This was collected by the Khoikhoi and used for medicinal purposes.

The gestation period is 225 days, which is remarkable for so small an animal. Two to five young are produced and become fully mobile soon after birth. After the first day they are largely on their own, returning to the mother to feed. Dassies multiply rapidly but are subject to predation by leopards, lynxes and eagles. As leopards have been exterminated on the Peninsula mountains, the return of the black eagle is most welcome for the control of the dassie population. Another interesting fact is that dassie populations appear to fluctuate greatly for reasons as yet unknown.

A few lynx or caracal still survive on the mountains, but as they tend to hunt domestic poultry on properties bordering the nature area they are not popular with landowners. The caracal is a handsome animal about the size of a fox-terrier, of a uniform reddish-fawn colour, with sharply pointed ears accentuated by tufts of black hair. The golden-amber eyes with dark blue pupils are a striking feature. It is a nocturnal animal of the open veld and a ruthless hunter of small antelope, dassies, hares, rodents and birds, ranging far and wide in search of prey.

Cape otters are still occasionally seen in the streams and intertidal zone of the Cape Peninsula Nature Area. The Cape clawless otter is dark brown with a lighter belly and white patches on the throat and chest. An adult weighs about 18 kg. The legs are short and the toes

Caracal. (*Photo: John Visser*)

Cape clawless otter. (*Photo: John Visser*)

are devoid of claws – hence their common name. The muscular tail, used in swimming, is slightly flattened. Although it is terrestrial the otter hunts in water and is usually found along rivers and vleis, but has also taken to inhabiting the coast, feeding on shellfish and crayfish. They are outstanding swimmers. The Cape otter is one of our most engaging and interesting animals. Should you by lucky chance come upon a family group, freeze in your tracks, and you may enjoy observing at first hand their skill in hunting and agility on land and in the water.

Smaller predatory animals which may be glimpsed while they are hunting in the dense undergrowth are the lesser or Cape grey mongoose, which is a dark pepper-and-salt in colour with a sharp snout, beady black eyes and a long bushy tail, and the small-spotted genet. These beautiful small catlike creatures have thick yellowish-grey fur marked with distinctive black patches.

Spotted genet (*Genetta* sp). (*Photo: Zelda Wahl*)

They are largely nocturnal, lying up by day among loose boulders at the foot of the cliffs or in the fork of a tree. They feed on rats, lizards and birds which they often hunt by scent. When cornered, the genet emits a nauseating musky smell.

Those who work regularly on the mountain frequently see feral domestic cats, which constitute a real problem in the management of any nature area. While they assist in controlling field rodents, they are also great predators of birds, whose diminishing numbers are a matter of concern. Packs of free-ranging dogs harass and may even kill antelope and ground-nesting birds.

The order Rodentia with over 6 000 species worldwide is the most successful mammalian group. The order includes the capybara, nutria, porcupines, beavers, squirrels, hares, rabbits and a large variety of rats and mice. Rodents are the best represented of the vertebrate fauna in the nature area today. Although differing widely in appearance, a characteristic feature of all rodents is the single pair of incisor teeth in the upper and lower jaws. These chisel-like teeth grow continuously and wear to a sharp cutting edge from the constant gnawing.

The porcupine, the name being derived from the French "porc espin", is the largest rodent in Africa and one of the largest in the world, weighing up to 22 kg. In addition to hair, the body is covered with stout sharp black and white quills and long thin spines which the animal is able to erect for defence when threatened. These sharp spines can inflict a painful wound and being barbed are difficult to withdraw. Porcupines are nocturnal, emerging from their burrows or caves to feed on roots, bulbs, corms, succulent vegetation and fruit. They often root up large areas in search of bulbs and can cause extensive damage in gardens. Being nocturnal

Porcupine. (*Photo: Cape Nature Conservation*)

they are seldom seen, but one frequently finds discarded quills on the mountain paths.

The Cape hare is the commonest species found in South Africa. In contrast to rabbits, hares shelter by day in forms under bushes or in dense tufts of grass. The ears are long and hearing acute. They are very fleet-footed, their only defence against predators.

In addition to the introduced brown and black rats, a variety of field rats and mice occur in the nature area, including gerbils, the long-tailed rat, striped rat, meadow rat, the vlei and bush otomys, tree mouse, fat mouse and multimammate mouse. They are all vegetarians living on a variety of plant material. Gerbils live in burrows, ranging out for food. The tree mouse nests in dense bushes, while most field rats and the otomys make intricate systems of runways in thick grass. They are largely nocturnal but are occasionally seen by day scuttling from shelter when disturbed. The multimammate mouse is very prolific, producing litters of up to twenty young after a gestation period of only 30 days. It is a carrier of the flea causing bubonic plague.

The Cape mole-rat (*Georychus*) is adapted to a subterranean life and reveals its presence by characteristic mounds of earth pushed up from its extensive system of burrows. Mole-rats are vegetarians feeding on roots, tubers and bulbs, often stored in underground chambers when food is plentiful.

In the sandy flats at the foot of the mountain one frequently sees the large mounds of the Cape dune mole (*Bathyergus*), which is endemic to the Cape. This is one of the largest moles, weighing up to 1 kg and with a body length of 30 cm. Having similar habits to the mole-rat, it can cause extensive damage in gardens. In the diaries of Governor Van Riebeeck one finds frequent reference to the problem of dune moles in the Company's gardens, and interesting descriptions of the traps which were devised to control them. Often confused with the mole-rat is the small, insectivorous Cape golden mole (*Chrysochloris*), which tunnels just under the surface of the soil in search of worms, insects and other invertebrates. Although it does not feed on plants, the raised tunnels in seed beds and lawns are the bane of gardeners.

The caves and sheltered ravines of the mountain are the habitat of a variety of bats, both insectivorous and fruit-eating. Bats are the only mammals to have achieved true flight. To this end the forelimbs have become lengthened and modified to support a thin but tough wing membrane. The hind limbs are short and provided with claws by means of which bats hang upside down while at rest. In general only one young is born after a gestation period of 70 days, and a remarkable fact is that all the females of a colony give birth within a day or two. The newborn young cling to the dense, soft fur of the female and are carried by her until they are strong enough to be left in the nursery cave.

The Cape fruit-bat has a body of 15 cm and a foxlike head with large ears. At dusk they descend on the gardens in the suburbs to feed on fruit and berries. They

Cape fruit-bat. (*Photo: John Visser*)

are particularly partial to the fruit of the loquat, guava, date palm and yellowwood trees. At dawn they return to communal roosts in the mountain. Only one young is produced at a time, and is carried by the mother on her foraging expeditions until it is strong enough to fend for itself.

At least seven species of bats occur in the Cape Peninsula, the commonest being the horseshoe bat, Schreiber's long-fingered bat, the long-tailed house bat, the Cape hairy bat, the Cape long-eared bat and the tomb bat. Towards dusk, these bats emerge from their roosts in caves or hollow trees to forage for food. On a calm evening they may be observed flying low, hunting for insects. They play a very important role in controlling insect populations.

Antelope, which are a characteristic feature of the fauna of Africa, are seldom seen on the mountains of the Cape Peninsula today, having been poached to extinction. Only a few remnants of the smaller species such as grysbok, grey reebok and an occasional steenbok or duiker may still be seen. The grysbok is a small antelope with a thick, bright rufous coat liberally sprinkled with long white hairs, especially on the back and haunches, giving it a grizzled appearance. The short horns are somewhat hidden by the large oval ears. Secretive by nature, grysbok inhabit dense scrub. They are territorial and their droppings are usually deposited on the same spot, known as a latrine patch.

In contrast to the grysbok, the steenbok makes a scrape in a sandy area to bury its droppings. This graceful antelope with a slender body, long legs, large ears and a sleek reddish-brown coat stands about 55 cm at the shoulder. The male carries short, sharp horns about 10 cm in length. Mainly a grazer, the steenbok is also a delicate browser of young leaves, shoots and flowers. It is a solitary animal, but may be found in pairs inhabiting an established territory.

The grey duiker is a larger animal standing over 60 cm at the shoulder and exhibiting considerable variation in colour, from dull grey to rufous, with a white belly and tailflash and a characteristic tuft of hair on the crown. There are conspicuous glandular orifices below the eyes, a morphological characteristic of duikers in general, their secretions being used to mark territories. Only the males carry horns. It is a skulking animal, seldom leaving cover. It feeds at night, grazing and browsing on vegetation and fruits.

On the more isolated plateaux of the southern Peninsula there are still a few grey reebok. This is a medium-sized antelope, light grey in colour with white underparts. Characteristic features are the long narrow ears and short bushy tail. The reebok is an antelope of the open veld and is usually seen in small groups. In flight it has a graceful rocking-horse-like action, clearing obstacles with ease while displaying the white flash under the tail. This serves as a warning signal to other members of the group. In the evenings, reebok often descend into the valleys to feed.

Grysbok.

Grey duiker. (*Photo: Cape Nature Conservation*)

The klipspringer is well named, being a characteristic antelope of rocky outcrops of our hills and mountains. Once plentiful, even on Table Mountain, it has been hunted to extinction in many areas. In contrast to other antelope, a klipspringer walks on tiptoe and its peg-like hooves are well adapted to ensure a firm footing as this small, agile antelope leaps from rock to rock. It is a sturdy animal, standing about 50 cm high at the shoulder, and the rounded body is covered with coarse grey bristle-like hairs which blend with the lichen-covered rocks of its habitat. The springy, bristle-like

A herd of grey reebok.

Cape mountain zebra. (*Photo: W Morsbach*)

hair is easily plucked and was used for stuffing saddles in the early days. They are inquisitive animals, usually seen standing stiff-legged on some vantage point surveying the surroundings, and consequently fall an easy prey to poachers. Efforts are being made to reintroduce klipspringers to the Peninsula mountains and it is hoped they will again be a common sight in years to come.

The bontebok, an antelope endemic to the Southwestern Cape which was hunted to near extinction, has been reintroduced to the Cape Point Nature Reserve where there is a thriving herd. It is a striking antelope with contrasting rich brown and white coloration. Characteristic are the white blaze, rump, underbelly and lower leg markings. The back is sloping and horns of up to 50 cm in length are carried by both sexes.

Another rare animal which has been successfully reestablished at Cape Point is the beautiful, once endangered Cape mountain zebra. It is smaller than the common Burchell's zebra and has a denser pattern of stripes. Shadow stripes are absent and the dark barrel stripes do not meet under the belly, which is white. The hair around the nose and mouth is black and there is a distinct dewlap – a flap or fold of skin under the throat.

These animals were once common in the mountainous regions of the Southern Cape but were ruthlessly hunted for their skins. By 1937 this species was on the verge of extinction when a small herd donated by the

Michau brothers from their farm Doornkloof was established in a national park near Cradock in the Cape Province. After initial setbacks, these animals have increased in numbers and been distributed to other nature reserves and parks within their natural distribution range.

No account of the vertebrate fauna of the Peninsula mountain range would be complete without mention of the alien species, some of which are so well established in the nature area that they are more commonly seen than the native fauna. These include fallow deer, Himalayan tahr and the American grey squirrel. The historic Groote Schuur Estate borders on the nature area and alien animals introduced on the estate have in the past invaded the mountain. A pair of tahr which escaped from Groote Schuur zoo in 1935 had populated the greater part of Table Mountain proper by 1970. A census conducted by the Mountain Club of South Africa placed the population at 800 to 1 000 animals. Control operations were subsequently instituted and more than 1 300 animals have been shot to date, but there are still about 100 tahr on the mountain.

The tahr is a sturdy mountain goat, slightly larger than a domestic goat. They are dark fawn in colour and the rams have a shaggy dark reddish-brown mane and strong recurved horns. Their hearing and eyesight are acute and being exceptionally agile, sure-footed creatures, they are particularly able to negotiate the rocky terrain with ease, finding footholds on the narrowest of ledges. Despite intensive hunting, it has proved almost impossible to eradicate these wary, elusive creatures.

Herds of common fallow deer and Japanese deer are to be seen on the lower slopes of the mountain on Rhodes Estate. They are tame and a great attraction to tourists. They never venture on the higher slopes of the mountain and consequently their numbers are easily controlled.

An unfortunate introduction to the Cape Peninsula area in 1900 by Cecil John Rhodes was the American grey squirrel (*Sciurus caroliniensis*). Here they soon became firmly established and spread further afield to Stellenbosch and Somerset West and beyond. Fortunately, they are confined to plantings of pine and oak trees, which provide both food and habitat. Although they are attractive and amusing animals, they are most destructive of bird-life and are a pest in established fruit and nut orchards. Early efforts to eradicate them were unsuccessful as being rodents they multiply rapidly.

Himalayan tahr.

Birds

South Africa is remarkably rich in bird-life, both as regards species and numbers. The visitor is likely to see more birds on the Peninsula mountains than other vertebrates. Most birds are seen at their best in the early morning or late afternoon. The following brief account describes these birds according to their habitat. Bird-life is richest in variety as one enters the fringes of forested areas. Climbing from the Newlands side, the first birds to be encountered as one moves up through the residential areas will be the common garden birds such as doves, thrushes, shrikes, white-eyes, colies (mouse-birds), sunbirds, and the ubiquitous, introduced European starling.

Four species of doves occur on the mountain. The small laughing dove has a pinkish head, neck and throat, shading to deeper pink on the back and chest which is speckled with cinnamon. The wing coverts are slate-blue shading to grey on the tail, while the belly and undertail coverts are greyish-white. It is usually seen on the ground searching for seeds. The call is a gentle descending cadence of notes.

The cinnamon dove of similar size is a more colourful bird. The upper body is a mixture of russet brown and iridescent green with cinnamon underparts and a whitish forehead, cheeks and throat. It is a shy forest bird and may be seen sitting motionless on the ground under shrubs or trees. It lives on berries, seeds and insects.

The Cape turtle-dove is a larger bird, paler in colour

than the laughing dove, and is characterised by a very distinctive black collar on the back of the neck and a white band on the outer tail feathers. Similar in appearance, but slightly larger, is the red-eyed dove. It also has a black half-collar but is more pinkish in colour and the tail feathers lack the white band of the turtle-dove.

In the woody ravines one may be fortunate to see the handsome rameron pigeon. Larger than the doves, it is a dark greyish-blue in colour with white spots on the chest and shoulders and has a distinctive yellow circle around the eyes, which match the beak and legs. It feeds on fruits and seeds and is very partial to the berries of the wild olive. The related rock pigeon is of similar size, but slaty grey in colour with white spots and a pink tinge to the plumage. The bare skin around the eyes and the legs is red. The rock pigeon is a strong flier, making lengthy trips to grainlands to feed and returning each evening to roost on ledges of the highest kranses, where it also nests. In recent years these birds have moved into the cities, roosting and nesting on the protected ledges of tall buildings, and they are rapidly adapting to a man-made environment.

Rameron pigeon. *(Photo: C J Uys)*

Common birds of the undergrowth on the approaches to the mountain slopes are the Cape thrush and the Cape robin. The Cape thrush is a dusky olive-brown with a speckled throat, dull orange breast and yellow legs and bill. It is usually seen scratching in the ground-litter for insects and worms, or fallen fruit. The Cape robin is a smaller, more attractive bird. The head, nape and wings are dusty grey and the chest, rump and tail are bright chestnut. Distinctive features are the white stripe extending from the bill over the eye, and its call. Its jerky, hopping movements and twitching of the tail are also characteristic. The song is attractive and heard at its best when two male birds are competing. The name "janfrederik" is derived from its call.

Another denizen of the undergrowth, particularly along stream beds, is the boubou shrike (southern boubou). The head, nape, back and tail are a glossy black, while the underside is white with cinnamon flanks. Although they are inquisitive, they are difficult to see as they remain concealed in the shrubbery. They associate in pairs and sing attractive duets, but these are not as loud or melodious as the songs of the more obtrusive bokmakierie. Like the boubou shrike, the bokmakierie also lives in bush and shrub but usually perches in the open. It is a handsome bird, the overall colour being olive-green with a bright yellow chest and black bib. The tail is dark, with a pronounced yellow fringe. They feed on invertebrates, small lizards and frogs.

The fiscal shrike, the commonest of the shrikes, is black above and white below with a distinct white V on its back formed by the folded scapulars. The rump and tail feathers are grey. The fiscal shrike is aggressive, disputing any invasion of its territory. They feed on insects, lizards and small birds which are often impaled on thorns or barbed-wire fences, hence the common name butcher-bird. It is an obtrusive bird, often seen perched on some vantage point, singing or imitating the calls of other birds. It is a great mimic with an amazing repertoire.

Other insect-eaters are the Cape wagtail, the small, attractive Cape batis, the dusky flycatcher and the paradise flycatcher with its striking breeding plumage. Related to the batis, the dusky flycatcher is a slightly larger but rather dull bird, with a greyish-brown plumage and pale grey chest. In contrast, the male paradise flycatcher is a striking, bright reddish chestnut, with long tail feathers. The underparts are grey and the crested head and neck are deep metallic blue with a marked wattle around the eyes. The nests of both the batis and the paradise flycatcher are works of art, being neat cups placed in the fork of a tree and built of bark and fibres camouflaged with lichens affixed to the nest with strands of spider web.

The Cape white-eye is a bird of thick shrubbery, trees and urban gardens. It is a small, dull greenish-yellow bird, which congregates in flocks out of the breeding season. The underparts are grey with a brighter yellow

Cape batis. (*Photo: W Nicol*)

Paradise flycatcher. (*Photo: W Nicol*)

throat and vent. The eye is surrounded by a distinct white ring, hence the name. A closely related species is found in Australia where it is called the silver-eye. A recent immigrant to New Zealand, it was appropriately named "little stranger" by the Maoris. White-eyes eat both fruit and insects and forage in small flocks, constantly calling while moving from branch to branch in search of plant lice and insects. They are most engaging birds and in gardens play an important role in controlling aphids and greenfly.

Fruit-eating birds frequenting wooded localities are the colies, bulbuls and redwing starlings. The redwing starling, a handsome bluish-black bird with a glossy sheen and a reddish-brown panel in the wing, has a pleasant, clear, mournful call, but makes a scolding noise while feeding. Redwing starlings usually roost on kranses, decending to the lower slopes and urban gardens to feed on a variety of fruits and berries. During the nesting season the flocks break up into breeding pairs and the young are reared on invertebrates and fruit. The parent birds will dive at human intruders near their nest.

Introduced to the Groote Schuur Estate by Cecil Rhodes in 1899, the European starling has spread throughout the coastal areas of the Cape from the Orange River to beyond East London. Feeding by day in small groups on fruit and insects, they gather in vast flocks towards evening at communal roosting sites in tall trees. They are aggressive birds well adapted to urban areas, feeding at city dumps or in gardens and nesting in hollow trees, holes or under the eaves of houses. It is a smaller bird than the redwing starling, the plumage of the male being blackish-brown, spotted with white, and having an iridescent green sheen. The female is a duller black, lacking iridescence, while young birds are a dull brown.

Three species of mousebirds (or colies) may be seen feeding on ripe fruits, berries or soft vegetation. They are brown birds with long tails and a marked crest. The name mousebird is derived from the hairlike appearance of the feathers on the head and breast as well as the habit of creeping around in dense vegetation. They are gregarious, associating in flocks of ten to fifty birds and maintaining contact by twittering calls. At night they sleep clustered together. Their flight is characterised by rapid wingbeats interspersed with gliding.

In contrast to the colies, the Cape bulbul is a solitary bird. The general colour is sooty brown. It has a marked crest, a white wattle around the eye and bright yellow undertail coverts. They are lively, noisy birds with a cheerful, liquid call and melodious song, but are not popular in orchards or vineyards, where they cause appreciable damage. A close relative is the sombre bulbul, an unobtrusive, dull, olive-green bird which skulks in the dense vegetation and is seldom seen. Its presence is signalled by the clear, liquid call which it utters freely. It is a most tantalising bird, for although one can get to within a few metres of where it is calling, it usually succeeds in keeping out of sight.

One of the largest birds of the dense bush which may occasionally be seen near water, is the white-browed coucal. This handsome bird has a black head, streaked with white, red eyes and long black tail coverts barred with white. The wings are chestnut. It feeds on insects, lizards, young birds and even small mammals. Its distinctive call consists of a series of liquid, flutelike notes in a descending scale, which has been described as sounding like water being poured from a bottle. It is claimed that this bird becomes very vocal before rain; hence the popular name of rainbird.

Another important group of birds found on the lower wooded slopes and in the ravines are the owls. Being nocturnal birds they are less evident during the day, but the spotted eagle owl is the species most likely to be seen. It is a large bird, greyish-brown in colour, mottled, with white on the chest. Elongated feathers on either side of the head form two distinctive earlike plumes. The large eyes are golden-yellow with black pupils. As the wing feathers of owls are soft, their flight is silent and the prey is usually unaware of the predator's strike. They feed on small rodents, bats and insects. Spotted eagle owls usually associate in pairs for life.

An interesting feature is that, while both parents hunt, only the female feeds the young. The male carries food to the fledglings, but does not react to their cries for food as the female does, tearing the prey into small portions. Consequently, if the female is killed the chicks usually die, often surrounded by food. Young owls may leave the nest and fall to the ground before they are able to fly, but the parents are usually in the vicinity and will continue to feed them until they are able to fend for themselves. Well-meaning people often pick up young owls in the mistaken belief that they have been abandoned, but they are better left undisturbed.

Spotted eagle owl. (*Photo: Zelda Wahl, Cape Nature Conservation*)

The smaller wood owl, which nests in hollow trees, is also found on the mountain. The general colour is dark brown, flecked with white. There are no ear tufts. Wood owls prey on small rodents, birds and frogs. Barn owls also occur on the mountain, usually near a suitable roosting cavity in a hollow tree, cliff or old building. They feed largely on small birds and rodents.

The cuckoos are birds of the wooded areas, usually first noticed by their calls. Three species occur on the mountain during the spring and summer months: the red-chested cuckoo; the smaller diederik cuckoo; and Klaas's cuckoo. The red-chested cuckoo or "piet-my-vrou", so named from its call, is a breeding migrant arriving from Central Africa in the early spring. It is a noisy bird and at times its loud repetitive call can be irritating, especially on moonlight nights. It is a dark, bluish-grey bird with buff underparts barred with black, a russet chest and dark, full, rounded tail spotted with white above and barred on the underside.

In the Western Province, the "piet-my-vrou" usually parasitises the nests of the Cape robin, and its single egg hatches in about 15 days. Shortly after hatching, the young cuckoo proceeds to eject the eggs or fledglings of its foster parents from the nest. This it does by crawling under an egg until it is balanced on its back and supported between the wings. It then straightens its legs and with a heave pitches the egg backwards out of the nest. Ornithologists claim that this is purely a reflex action, a response to the presence of other objects in the nest, and that it only occurs within the first few days of hatching. Having cleared the nest of its stepbrothers and stepsisters, it now enjoys the full attention of both foster-parents and grows rapidly. It seems incredible that birds will feed a chick so much larger than themselves and so different in appearance from their own young.

Other migrant cuckoos found in the Cape are the smaller, iridescent, bronze-green diederik cuckoo and the somewhat similar Klaas's cuckoo, the former being distinguished by its barred tail and white eyebrows, white spots along the wings and outer tail feathers. Both species feed largely on caterpillars. The diederik cuckoo parasitises the nests of weaver-birds, sparrows and wagtails, while the Klaas's cuckoo lays in the nest of flycatchers and sunbirds. Once the chick is fledged, it joins the parent birds for a few weeks.

In the Cape Peninsula the wild peach (*Kiggelaria*) is often so heavily infested with the caterpillars of the common orange and black garden butterfly (*Acraea horta*) that the trees are almost defoliated. Cuckoos appear to be the only birds that feed on these repulsive-looking, dark brown caterpillars. It is interesting to watch an

adult diederik tenderising the caterpillars on a branch before feeding them to its fully fledged young. The male diederik will also feed the female.

Klaas's cuckoo is a brilliant glossy green with a bronze sheen, while the ventral surface is white. There is a distinctive green half-collar on the neck and upper breast, and the outer tail feathers are white, barred with green. The female is duller than the male, with bronze tones on the head, neck and upper mantle. Its plaintive call, "meitjie-meitjie", is uttered by the male only.

Along the mountain streams, in the ravines and particularly in Disa Kloof on the Hout Bay side of the mountain one may hear the raucous call of the giant kingfisher. The giant kingfisher is a striking bird. The back and wings are bluish-black with white spots, while the underside is white with black spots and the chest rufous. The heavy head has a pronounced crest, and the beak is large and sharp. The bird hunts from a convenient perch overlooking a likely pool, diving on fish and frogs which are carried back to its perch, tenderised and swallowed whole. It is particularly fond of freshwater crabs, which are repeatedly hammered on a convenient stone to shatter the shell and then swallowed piecemeal.

On slender branches overhanging a stream one may see the oval nests of Cape weavers. The male weaver has a bright yellow breast and olive-green back, while the female is a duller brownish-green. They live in colonies, building their intricate nests, which are works of art, from strips of grass or reeds. Having selected a suitable branch, the male removes the leaves. Strips of reed are then firmly attached to the end of this branch and formed into a ring. On this foundation the male weaves a sturdy oval nest, a remarkable feat when one considers that these are built purely by instinct and with beak and claws only. The male appears extremely proud of his efforts and alternates periods of intense activity with song. Hanging upside down from the structure, feathers puffed out, wings fluttering and "swizzling", he endeavours to attract a female.

Klaas's cuckoo. (*Photo: C J Uys*)

The female examines the nests under construction, and should she find one to her liking, will line it with soft grass and feathers and lay her pale blue eggs therein. She incubates the eggs and feeds the young while the male displays in the vicinity. He may even build another nest and try to attract a second female. Should his nest be rejected by the females, the male will tear it apart and build another. Some males have been observed to build as many as ten nests in succession, and it would appear that they become progressively more frustrated, for each new effort is poorer than the previous one.

In Orange Kloof one may occasionally also see a large brown bird with black legs, a heavy black beak and a distinctive crest: the hamerkop. It feeds on frogs, fish, crabs and aquatic insects which it catches by standing motionless at the water's edge until the prey comes within striking distance of its sharp bill. The hamerkop, *Uthekwana*, is immortalised in Xhosa folklore due to this habit and like Narcissus of Greek mythology is be-

Hamerkop. (*Photo: Izak Barnard*)

lieved to be admiring its own reflection. The hamerkop builds a huge nest of driftwood on a ledge or in a tree overhanging water. The entrance is on the underside, usually inaccessible to predators. The nest is used year after year by the parent birds, who renovate it prior to the breeding season. Unfortunately this bird is becoming rare due to nest piracy by Egyptian geese, owls and the introduced grey squirrel.

A number of so-called "game birds" occur on the lower slopes of the mountain chain, usually in the vicinity of plantations. The Cape francolin is endemic to the South-western Cape. It is a large bird, dull brown in colour and marked with white. The bill is red and the legs orange-red. It is a ground bird, flying only to escape predators or to move from one feeding-ground to another. They feed on seeds, bulbs, berries and insects. The smaller grey-winged francolin, commonly known as the partridge, may occasionally be seen, but these birds prefer the sandveld and renosterbos areas.

The helmeted guineafowl, although indigenous to southern Africa, was introduced to the mountain from the Groote Schuur Estate and is now well established. The general colour is dark greyish-blue, liberally speckled with white. A distinguishing feature is the horny red casque on the head and the bare blue neck and throat. They move around in flocks, spend most of the day feeding on the ground, but roost in tall trees at nightfall. Birds pair off in the breeding season, but members of the flock assist in herding and rearing the chicks.

In the more open fynbos of the lower slopes one finds the Cape sugarbird and a number of sunbird species which live in close association with the proteas and heaths. The Cape sugarbird is brownish-grey in colour with a yellow vent, white stripe below the eye and a long, curved beak for probing the protea blooms to obtain nectar. It also feeds on insects. The long tail feathers are a striking feature and are shown to advantage in courtship displays when the birds are vocal, uttering happy chortling and swizzling calls. This attractive bird is restricted entirely to protea veld and is endemic to the Cape.

Three species of sunbirds are to be seen on the mountain slopes, the handsome orange-breasted and lesser double-collared being usually associated with the fields of ericas, while the striking metallic green malachite sunbird is attracted to nectar-producing flowers. The male orange-breasted sunbird with a bright orange breast, marked with a band of metallic violet, has a shiny dark green head, throat and mantle. They feed on nectar, insects and spiders. The nest, placed a few feet above the ground in a dense shrub, is a neat bag of grass and fibres bound with cobweb and lined with softer material.

The lesser double-collared sunbird is recognised by its bright, metallic green head and neck and bands of blue and scarlet on the breast, while the rest of the body is ash grey. The female is a dull ash grey with darker wings and tail. The call is a characteristic high-pitched "cheep-cheep". During the mating season, the brilliant green malachite sunbird develops two bright yellow pectoral tufts, while the female is a drab yellowish-grey which has the advantage of rendering her less conspicuous while incubating.

Common but less colourful birds seen on the open fynbos slopes are the neddicky, Levaillant's cisticola, the Karoo prinia and the Cape grassbird. The neddicky is a small brownish-grey bird with a chestnut nape and crown. During the breeding season the male is very vocal, calling from a favourite perch. Levaillant's cisticola is one of the many species of cisticola found in South Africa. It is an attractive brown bird with a russet crown and distinctive black markings on the back and wing coverts. It is usually seen flitting above the vegetation of marshy areas, uttering a loud "tee-tee-tee" alarm call.

The Karoo or spotted prinia is about the same size, but is dull brown with a lighter breast marked with black streaks. They are excitable birds, uttering loud scolding calls when disturbed. The grassbird is a larger, more striking rusty-brown bird marked with chestnut and black and with a long pointed chestnut tail. The Knysna scrub warbler is a shy bird inhabiting thickets, especially of brambles. It is a uniform dull brown, but has an attractive sustained warbling song ending in a bubbling trill and is particularly vocal in the spring months.

Other songsters which may be encountered on the mountain trails are the Cape bunting and the Cape canary. The Cape bunting is a small brownish-grey bird

Cape bunting.

with grey crown and back streaked with black, grey underparts and chestnut wings. The head has distinctive black and white stripes. It is a fairly tame bird, often seen perched on a rock while singing a canarylike song. It hops in sparrowlike fashion, feeding on seeds and insects.

The Cape canary is a bird of similar build but greenish-yellow in colour, with brown streaks on the back and wings and a bluish-grey nape. It is usually seen in flocks adjacent to scrub and forest areas. This is one of our best songsters, and hearing a flock of Cape canaries in full song is a never-to-be-forgotten experience.

Moving further up the mountain, one encounters the birds associated with rocky outcrops and boulder-strewn hill slopes such as the ground woodpecker, the Cape rock thrush and the familiar chat. The ground woodpecker, endemic to the Cape, is remarkable in that it lives on open mountain slopes and is frequently seen in pairs or family groups perched on a rocky outcrop. It is a brown bird with a pinkish-red breast and belly and a bright red rump. It feeds on insects trapped with its sticky tongue.

Ground woodpecker. (*Photo: C J Uys*)

The Cape rock thrush is a reddish-brown bird with darker markings on the flight feathers and tail. A distinguishing feature of the male is the slate-blue head and neck. It has a characteristic habit of flicking its wings after landing. It is catholic in its diet, feeding on worms, insects, molluscs, frogs, seeds and berries.

Another rock-lover is the Cape siskin, a charming little bird, dark in colour with a yellowish-brown breast and white tips on the wings and tail. It associates in small parties, feeding largely on seeds and insects. The familiar chat is also a brown bird with russet ear coverts and underparts. The tail and rump are deep russet. It is a lively bird, becoming quite tame if unmolested, and will sometimes approach picnic parties to scavenge for crumbs.

The upper cliffs of the mountain are the habitat of a variety of birds, from eagles and hawks, to swallows, swifts and martins. There was much rejoicing among conservationists recently when a pair of black eagles was seen on the mountain after an absence of many years, and actually nested. This magnificent bird is usually observed soaring high over the cliffs in search of dassies, which are its major prey. It also feeds on small antelope, young baboons and guineafowl. The plumage is shiny black with a distinctive V-shaped white marking on the rump. In flight the white bases of the primary feathers are visible.

Near Cape Point one may see a pair of fish eagles. The fish eagle is a dark chestnut-brown with a distinctive white head, neck and breast and a yellow beak. This handsome bird, usually found on estuaries, is characterised by a wild, eerie call, and feeds largely on fish which are taken in a swoop while basking near the surface. Reduced almost to extinction, these striking birds are slowly making a comeback due to the strict legal protection of all eagles.

Two species of buzzards are frequently seen on the mountain. The first is the jackal buzzard, with dark slate-grey plumage marked with white flecks, and chestnut chest and tail. The name is derived from its call, which is reminiscent of the cry of a jackal. The steppe buzzard is a summer migrant from Europe and North Africa and is a brown colour, barred and blotched with

Fish eagle with prey. (*Photo: Izak Barnard*)

white. Steppe buzzards feed mainly on insects but will also take small birds, lizards and carrion at times.

The smaller hawks include the rock kestrel, the red-breasted sparrowhawk and the black-shouldered kite. The rock or common kestrel is deep rufous, spotted with black, and has a bluish-grey head and tail, ending in a broad black subterminal band. They feed on small mammals, insects and birds and usually nest on ledges. The red-breasted sparrowhawk has a slate-grey head and back, and plain rufous underparts. The tail is broadly barred with black and white. The legs, eyes and cere are a distinctive yellow. They feed largely on small birds and insects which they catch in flight, having developed the amazing skill of flying at high speed through woodlands.

Another raptor of the woodlands is the African goshawk. It is a dark slate-blue hawk with white underparts barred with brown, and yellow legs. An unobtrusive bird, it keeps to the dense forest canopy except in the early mornings or evenings when it flies high above the trees uttering high-pitched calls. It hunts in the forest, usually taking its prey – which consists of birds, small mammals, reptiles and frogs – in flight.

The attractive black-shouldered kite, usually seen in pairs, is probably the commonest bird of prey on the mountain. The general colouring is pale grey above with black shoulder-tabs and a white head, neck and chest. The legs and claws are bright yellow and the eye red. The flight resembles that of a gull, but it frequently hovers while seeking its prey of field rodents, lizards or insects.

The white-necked raven uses the precipitous cliffs for nesting and roosting, but ranges far and wide in search of food. It is a large, shiny black bird characterised by a white band around the nape of the neck and a heavy bill. The raven is omnivorous but is particularly attracted to carrion. They are regarded as pests by farmers as they feed on eggs and young birds and will peck at the eyes of sickly or newborn lambs, thereby killing them. They kill small tortoises by taking them aloft and dropping them onto rocks to shatter the carapace in order to feed on the flesh. The raven is an intelligent bird and is still reasonably plentiful, despite persecution by some landowners.

A group of birds usually associated with the cliffs and rock-faces on the mountain, particularly during the breeding season, but also found in other habitats are the swifts, swallows and martins. The layman often finds difficulty in distinguishing between these birds because they are usually only seen in flight. Being aerial-feeding birds which live largely on insects, they have long pointed wings and, in some species, forked tails, wide gapes and small feet.

Swifts spend most of the time in the air and are therefore more highly adapted to an aerial life than swallows. In some swifts the feet are so poorly developed that they can hardly walk. The martins are dull brown birds and are weaker fliers than the swifts or the swallows. All three groups are gregarious. They congregate in large flocks at the start of the Cape winter before migrating northward.

Three species of swift are frequently seen in the Peninsula, all being summer migrants. The black swift has a black body with a bronze sheen on the wings and a pale whitish throat. It breeds on precipitous rock-faces and the nest, made of rootlets and feathers cemented with saliva, is placed in an appropriate crevice. Although based on the mountain, these birds range far out over the Cape Flats to feed. Similar in habits to the black swift is the Alpine swift, dull black in colour with a white chin and belly. It breeds on mountain cliffs and also ranges far afield for food.

The white-rumped swift is the smallest species, almost completely black in colour except for a narrow white band on the rump. Distinguishing features are the long, deeply forked tail and the rather harsh shrieking calls uttered in flight. Although this species also utilises cliffs for breeding, it usually nests under the eaves of buildings, often appropriating the nests of the stripe-breasted swallow. In fact, swifts will even eject the rightful owners from their nest, destroying eggs or chicks in the process, and due to this aggressive habit are not popular.

While by no means confined to the mountain, three species of swallows found there are the greater striped swallow, the white-throated swallow and the well-loved European swallow. All three species are summer migrants, only the first two breeding at the Cape. The greater striped swallow is the largest of these swallows, being recognised by a reddish chestnut head, pale chestnut rump, creamy striped breast and forked tail. The nest, made of blobs of mud placed under the overhang of a cave or eaves of buildings, is gourd-shaped, with a long tubular entrance. These nests are often pirated by white-rumped swifts.

The white-throated swallow, a smaller bird, has a shiny bluish back and white underparts, with a black band on the lower throat. The forehead is reddish-chestnut and the tail shows white windows in flight. The flight is much swifter than that of the striped swallow. The cup-shaped nest of mud pellets is usually placed under an overhang in the vicinity of water. Although it does not nest in South Africa, the European swallow is the most numerous of our swallows. They nest in Britain and Russia, arriving here in early October and leaving again by March. During the weeks prior to departure they gather in large flocks. The European swallow resembles the white-throated, but is slightly larger. A distinguishing feature is its reddish-brown throat. Both birds have the black band on the lower throat, but this is much narrower in the white-throated species.

The martins are a group of greyish-brown birds, closely related to the swallows and similar in build and habits. The rock martin is a resident species and, as the name indicates, needs cliffs for nesting but has also adapted to using buildings for this purpose. The nest is cup-shaped, placed under an overhang or under the eaves.

Returning home towards dusk along the contour path above Kirstenbosch, one might be so fortunate as to hear the distinctive whistling call of a fiery-necked nightjar. The call has been set to words as "Good Lord – deliver us"; hence the name of litany bird. Nightjars nest on the ground, where their general brown colour, barred and spotted with darker brown, black and white provides a perfect camouflage among the ground litter of leaves and bark. The large eyes are reddish-brown and the feathers on the neck and upper chest are rufous. While usually solitary, nightjars also associate in pairs. They are insect feeders, hawking them at night from a convenient perch.

At its southern end the Cape Peninsula Nature Area adjoins the Cape of Good Hope Nature Reserve, where the habitat differs markedly from that of the mountains. The reserve is far more varied in topography, for in addition to small mountains it includes areas of open flats with temporary vleis and a coastline of dune veld, beaches and rocky headlands.

The fauna and flora of this interesting reserve has been described elsewhere (Hey, 1987). Suffice it to mention here that the bird-life of the reserve includes a number of species not usually found on the mountain, such as larks, the longclaw, plovers, herons, ducks, gulls, sandpipers, cormorants, terns and oystercatchers, an occasional penguin, and many others. Here one can also see Cape mountain zebra, bontebok and other large antelope introduced to the reserve, the rare Cape otter, seals, dolphins and, in early spring, right whales, lazing in the shallows.

Cape jackass penguins. (*Photo: Cape Nature Conservation*)

Reptiles

South Africa, and particularly the Cape Province, is renowned for the richness and diversity of its reptilian fauna. It has been estimated that about nineteen species of lizards, 22 species of snakes and three land tortoises are to be found in the Cape Peninsula, the majority of which also occur on the Peninsula mountains.

SNAKES

Many people are terrified of snakes, and in the case of a few species, justifiably so. But the majority of snakes are harmless, and even the poisonous species avoid man and will only attack when threatened or trodden on. While out walking, one should therefore look carefully before stepping over boulders or logs in the pathway, for a puffadder could well be lying there. When suddenly encountering a snake, stand perfectly still and it will usually move off.

The puffadder is a common species and is responsible for the majority of snake-bite casualties. It strikes with remarkable speed and is very poisonous, the venom being cytotoxic. This flat-headed, thick-bodied, sluggish snake, usually less than 1 m in length, derives its name from the habit of inflating its body and giving a loud, threatening hiss. Its colour is dark brown, strikingly marked with yellow and black, being particularly bright after sloughing off its old skin. Also found on the mountain is the smaller bergadder, with distinctive triangular black markings and two silvery-white dorsolateral stripes running the length of the body. It is irascible when disturbed, and is a danger to the unwary rock climber as it lies basking on a ledge.

Other dangerous snakes include the Cape cobra, the rinkhals and, to a far lesser extent, the boomslang. The Cape cobra is a large, slender snake about 2 m long, dull yellow or brown in colour, with the characteristic habit of rearing up when threatened and spreading its hood in typical cobra fashion. While they prefer to escape, they will attack if cornered. The venom is neurotoxic and may result in death unless treatment is applied almost immediately.

A close relative is the rinkhals, sometimes called the

Puffadder. (*Photo: G MacLachlan*)

Boomslang. (*Photo: John Visser*)

ring-necked spitting cobra, owing to its habit of rearing up with expanded hood and "spitting" venom when threatened. It does not really "spit", however, but ejects the venom from the poison fangs in the form of a thin spray up to a distance of 2,5 m. Should the droplets strike the eyes they may cause temporary blindness. The pain is excruciating and the eyes should be bathed as soon as possible with a weak solution of antivenomous serum. If this is not available, milk, water or any bland solution may be used. The rinkhals is black in colour with two distinctive white or yellowish bands on the throat. This cobra has the habit of turning on its back and feigning death should its attack fail. It should not be handled, however, as it can strike unexpectedly.

The boomslang, as the name implies, is a tree snake, its slender body being specially adapted for gliding through the branches of trees and shrubs. It varies in colour from dark green to a rich brown with black markings. The boomslang is not aggressive, in fact it is one of the least aggressive of snakes and will only bite if provoked. The haematoxic venom is deadly, but as the fangs are small and situated far back in the mouth, the bite is seldom fatal.

Non-poisonous snakes which may been sunning themselves in sheltered spots include the mole-snake, the whip-snake, the cross-marked grass snake and the rhombic skaapsteker. The mole-snake is one of our most useful snakes, living largely on moles and other rodents. It is one of the few predators that can pursue them into their burrows and being non-poisonous, kills its prey by constriction. Adult mole-snakes are a uniform dark brown to black and often grow to a length of over 2 m. In the words of that doyen of naturalists, Dr Walter Rose: "Harmless as earthworms and beautiful as if carved by some super artist from flexible sentient ebony, the mole snake's claim to our protection is surpassed by no other South African animal."

The whip-snake and cross-marked grass snake are closely related species and completely harmless to man. They are slender snakes, about 1 m long and marked with longitudinal stripes or a series of spots. They are very active, moving through the vegetation at amazing speed when alarmed or when pursuing their prey, which consists largely of lizards. The skaapsteker, although poisonous, is of gentle disposition and harmless to man. The name is misleading as it is incapable of killing sheep or other large animals and feeds on mice, lizards and frogs. The colour is light brown above with a conspicuous pattern of black markings and a whitish underbelly with spots and blotches. It attains a length of 85 cm.

Other harmless snakes frequently found in damp localities on the mountain are the red-lipped snake, the russet garden snake and the Cape many-spotted snake. The Cape many-spotted snake is brown with a green tinge and a longitudinal stripe of dark brown spots and whitish speckles. The underside is dull blue to green. It may grow to 50 cm in length and feeds largely on frogs.

The common slug-eater (*Duberria*), also known as the russet garden snake, seldom exceeds 50 cm and varies in colour from brown to bright russet, with a broken black

Russet garden snake. (*Photo: G MacLachlan*)

vertebral stripe. The undersurface is yellow. Completely harmless, it has the habit of rolling itself into a spiral when threatened, hence the apt Afrikaans name of "tabakrolletjie". Its diet consists almost exclusively of slugs and thin-shelled snails and it is capable of extracting snails from their shells.

The red-lipped or herald snake is olive-brown, marked with small white spots. The underparts are creamy white. A characteristic feature is the bright orange or red upper lip, hence the name. It is also found in the Eastern Cape, where it was named after the *Eastern Province Herald* by its discoverer. The average length is 75 cm but the occasional specimen may grow to over 1 m. This species is a nocturnal hunter, feeding largely on frogs. It is a harmless snake, but when molested it puts on a show of aggressiveness, striking out with mouth agape.

In the vicinity of the mountain streams one may occasionally see a water snake. These harmless snakes are only partly aquatic, but live near water. They are slender, usually about 60-75 cm long, and olive-brown, with a pearly-cream ventral surface. They feed on frogs, tadpoles and fish which are caught underwater and carried ashore to be swallowed.

While the risk of being bitten by a snake on the mountain is minimal, hikers are advised to wear strong boots, long trousers or thick socks to protect their ankles. In the unlikely event of someone being bitten, the victim must rest immediately and be kept warm and comfortable while help is sought. The only satisfactory treatment is the injection of the correct antiserum, and for this reason it is essential that the snake be identified. As a temporary measure, a pressure bandage may be applied over the bite and if possible over the entire limb, to reduce the spread of the poison. Tourniquets should not be applied by amateurs as severe tissue damage may result.

LIZARDS

A variety of lizards are to be seen on the mountain, sunning themselves on the lichen-covered rocks. None of them is poisonous; in fact, there are no poisonous lizards anywhere in South Africa. The only two poisonous species both occur in America.

The crag lizard is the largest local species, averaging 30 cm in length. It inhabits the highest kranses, taking refuge in cracks and crevices when disturbed. The male is more brightly coloured than the female, being black with thin bright yellow stripes. The belly is a dirty grey, while the legs and tail are bright yellow underneath. The zonures, or girdle-tailed lizards, are smaller rock-dwelling relatives with similar habits. They are covered with characteristic serrated shields which are particularly well developed on the tail, giving it a spiky appearance. They seldom exceed 18 cm. Other common rock-dwellers are the agile agamas or koggelmanders – the name being derived from their habit of bobbing the head while sunbathing. They are greyish in colour while the heads of the males are often bright blue.

Koggelmander. (*Photo: Dudley D'Ewes*)

The skinks are one of the largest families of lizards found in South Africa. These smooth, shiny lizards are represented on the Cape Peninsula mountains by two species, the Cape speckled skink and the larger Cape three-striped skink, which seldom exceeds 22 cm. Although skinks have relatively short limbs, and some species are actually limbless, these lizards can move surprisingly fast when threatened. They are dark brown in colour with distinctive longitudinal markings and a natural sheen, and shelter under rocks or in dense vegetation.

The common or true lizards are well represented in the Cape, three species being found on the Cape Peninsula. The best known of these is the ocellated sand lizard which has a somewhat rough, greyish-white back, with large eye spots often forming a definite stripe. It is a very agile lizard, darting out from cover to snatch an unwary insect. It is not a true mountain species, being confined to the lower sandy areas.

One of the most extraordinary lizards is the whip or snake lizard. This interesting creature is often mistaken for a snake, due to its slender, serpentine form and tiny, rudimentary limbs. Nevertheless, it possesses all the typical characteristics of a lizard and belongs to the same family as the girdled lizard. Whip lizards attain a length of about 48 cm and inhabit the grass-covered slopes and plateau areas.

The geckos are a very ancient family of lizards, and by far the largest in South Africa. The digits in most species are dilated into adhesive pads which enable them to negotiate smooth surfaces with ease. Normally geckos live under stones or in crevices, emerging only at dusk. They harmonise so well with their surroundings that they are difficult to locate. Some species are able to change colour to suit their surroundings. They tend to enter houses, where they may be seen on the ceilings at night, hunting insects attracted by the light. Geckos are completely harmless, timid creatures, yet they are held in superstitious dread by some people. Their sole defence when threatened is to shed their tail and scurry off, leaving a wriggling tail to distract the predator. The marbled gecko and the ocellate gecko are the two species found on the Cape Peninsula.

The chameleons are a distinctive family of lizards, inhabiting creepers and dense shrubs which attract insects, their chosen food. These are caught by shooting out the long, sticky tongue which can be extended to almost the length of the body. The eyes are large and can move independently, while they also have sharp binocular vision essential for catching their prey. An amazing characteristic of chameleons is their ability to change colour to blend with their surroundings. The species found in the Cape Peninsula is the Cape dwarf chameleon, a member of a genus which produces its young alive, in contrast to the other species, which lay eggs. The Cape dwarf chameleon is not a true mountain species, being found in the lower areas, especially in the vicinity of vleis.

Cape chameleon. *(Photo: Dudley D'Ewes)*

TORTOISES

The old Cape Province has the richest diversity of tortoises in the world. Twelve species occur within this region, of which three are endemic. Unfortunately, only the occasional tortoise may still be seen on the lower slopes of the mountain as their numbers have been sadly reduced over the past 300 years. Being slow-moving creatures they are usually killed or badly scorched by veld fires which have ravaged the Peninsula mountains for centuries.

Of the eleven species of land tortoises which occur in South Africa, only two are found within the Cape Peninsula Nature Area. The angulate tortoise is largely restricted to the coastal fynbos. It is a relatively small animal with a shell length of about 20 cm. The carapace is elongated and dull yellowish-brown in colour, with black margins to the shields and black triangular markings around the edges. These tortoises remain active throughout the year, as do most South African tortoises.

The second land tortoise found on the Peninsula is the Cape or southern padloper, about 12 cm long. The carapace is slightly flattened dorsoventrally, and this tortoise can be distinguished by the fact that it has only four claws on each forefoot. The overall colour of the carapace is olive-brown or dull green, with depressed reddish-brown centres to the shields. This species also occurs on the dry lower slopes. Mountain tortoises are occasionally encountered in the nature area. However, these have been introduced by well-meaning but misguided persons.

In permanent waters such as Sirkelsvlei near Cape Point, numbers of Cape terrapins or water tortoises can be seen basking on emergent rocks or sandbanks. This is the only species of water tortoise in the Cape and may attain a weight of over 2 kg. The flattened, smooth shell of these water-dwellers is a dull blackish-brown, and they have long necks, a pronounced tail and a broad head. The hind toes are webbed. They are largely carnivores, feeding on frogs, tadpoles, fish and the occasional duckling, although they may also consume vegetation. They are known to move overland from one water body to another and during times of drought will burrow underground.

Tortoises have inhabited this planet for over 200 million years, and it would be a sad indictment of our stewardship of nature if we were to cause them to become extinct. While there are natural predators of tortoises there can be no doubt that their depletion is due to man's activities. From prehistoric times tortoises have been popular as food for indigenous peoples, vast areas of tortoise habitat have been converted to agricultural land, and our mountains are ravaged periodically by veld fires. Until recently, tortoises were exported to Europe in large numbers for the pet trade. Our nature areas and national parks may well become the last stronghold of our diverse tortoise population, and it is consequently essential that these be enlarged and managed accordingly.

The amphibians

In a most readable book on reptiles and amphibia, published in 1950, Dr Walter Rose stated that nineteen species of frogs and toads were to be found within a few kilometers of the centre of the city of Cape Town – a remarkable fact when one considers the long, dry summers experienced here. While habitat destruction is reducing their numbers at an alarming rate, many of the species are still to be found in this area, about ten of them occurring on the mountains. Being small and unobtrusive, however, amphibians are usually only seen by those with the interest and patience to search for them.

The spotted grass frog (*Strongylopus grayi*) advertises its presence by the chorus of soft, click-like calls which one frequently hears when traversing a swampy area. They are mottled, brownish-grey in colour, and trying to locate one hidden among tufts of reeds or swamp grass can be a frustrating experience as they are cunning ventriloquists. Should one approach too closely, they stop calling, which makes spotting them even more difficult. This frog differs from the closely related Cape river frog, in that it lays its eggs in damp hollows and not in water. The developing tadpole remains within the gelatinous capsule to emerge with the first heavy rains and be washed into a pool of water for further development.

A larger relative is the long-toed or Cape grass frog (*Strongylopus bonaespei*), which mainly inhabits the seepage areas of the plateaux. It is characterised by the markedly bright silver-yellow or brown dorsal stripe, sharp pointed snout, and relatively long hind legs. A champion jumper, it can cover over 2 m in a single leap. The Cape river frog (*Rana fuscigula*) is a much larger creature of up to 12 cm in body length. It is aquatic by nature and consequently the hind toes are webbed. It is usually seen sitting beside a pool into which it leaps with a characteristic "plop" when disturbed.

Two very interesting species of frogs live in the cool, shady mountain ravines. *Arthroleptella*, the chirping frog, only about 2 cm in length, inhabits the moist, moss-covered sides of the ravines. They can be located by their soft chirping calls. The eggs, encased in a jelly-like mass, are laid in the damp moss. The tadpoles

Ghost frog (*Heleophryne rosei*). (*Photo: A de Villiers*)

squirm about in the moss and rapidly metamorphose into frogs. This species, which is only found in the South-western Cape, is of great scientific interest.

In the permanent, swift-flowing, peat-stained mountain streams lives the remarkable Table Mountain ghost frog (*Heleophryne rosei*). This rare frog is found only in certain ravines on Table Mountain and has the smallest distribution of any South African amphibian. It is a secretive animal, hiding under stones along a watercourse. The colour is dull green with reddish reticulation and a white semi-transparent underbelly revealing the viscera; hence the name ghost frog, which also relates to its habitat. The ends of the toes are expanded into suction discs enabling the animal to climb around on slippery rock-faces and cling to the rocks of the swift-flowing streams it inhabits.

Although the adult frogs are seldom seen and were, in fact, only recently identified, the wedge-shaped, yellowish-brown tadpoles are commonly seen in rock pools. A characteristic feature of the tadpoles, which are adapted to life in these swift-flowing, acidic brown waters, is the large sucker-like mouth used effectively to cling to the stones on the stream bottom while grazing on algae.

A purely aquatic frog found in the vleis of the Cape Peninsula is the platanna or clawed toad (*Xenopus laevis*). The platanna is one of our most interesting and valuable amphibians. The scientific name *Xenopus* means "strange foot", only the hind foot being webbed and the three inner toes provided with sharp claws. The forelimbs are poorly developed and used largely for feeding. Platannas are entirely aquatic but must periodically rise to the surface to breathe, although they also respire through the skin. They are largely predators, feeding on aquatic creatures they can catch and stuff into the large mouth. They also scavenge and are consequently easily caught in cylindrical wire traps with a cone-shaped entrance and baited with offal. The teeth are small, and the animal lacks a tongue.

For many years scientists believed that there was only one species of platanna, namely *X. laevis*. Dr Walter Rose, who devoted years of study to the amphibians of the Western Cape, had questioned this and eventually became convinced that there was another species, similar in appearance but smaller, shyer and less aggressive. This he named *X. gilli* after Dr E L Gill, a director of the South African Museum. He called this species the sago-belly platanna due to the dark, reticulated markings on the ventral surface. He advanced the theory that this was an older species endemic to the Cape, which is gradually being replaced by the more aggressive *X. laevis* with which, incidentally, it will breed. Today, *X. gilli* is recognised as a species, but there is concern that it is becoming rare, if not endangered, and steps are being taken to ensure its conservation. A small vlei in the Cape Point Nature Reserve has been enclosed to serve as a breeding sanctuary for this species.

The breeding season of *Xenopus* extends from late winter through into midsummer, provided there is a plentiful supply of food for the developing tadpoles. The eggs, which are laid singly and not in strings of jelly as with the land toads, are sticky and adhere to the stems of aquatic vegetation. After hatching, the larvae remain attached to the plants by mucous threads until the gills are fully developed and the mouths open. Platanna larvae are dorsoventrally flattened and provided with long whiplike tails. They are almost transparent, except for the abdomen which has a silvery coating. They remain in shoals, hanging head down, with the tip of the tail flagellating rapidly, while they feed on algae and infusoria. As they grow, they feed on larger items. The development is similar to that of other frogs, the hind legs appearing first. In contrast to most amphibia they remain aquatic when fully developed.

A small frog inhabiting the low-lying swampy areas is the graceful arum lily frog (*Hyperolius horstockii*), barely 4 cm in length. They have the ability to change colour from light cream to dark brown, depending upon the light, temperature and background. They are often found sitting motionless in an arum lily bloom, in wait for insects, the legs neatly tucked under the body. The fingers and toes are webbed and end in suction discs, enabling the creatures to cling to the thin stems of reeds. They are extremely agile and execute amazing leaps from plant to plant.

Toads are common in the Western Cape and differ from frogs in having a dry, rough skin richly provided with glands which secrete a white mucus, an effective defence mechanism which causes many predators to avoid them. Toads live on land, resting by day in some cool retreat and emerging at dusk to hunt insects and other invertebrates. The legs are shorter in relation to the size of the body than those of frogs, and consequently a toad moves by means of short hops or by crawling. Toads have large, golden-yellow eyes with black pupils.

During late winter or early spring, toads often congregate around semi-permanent pools to mate and spawn. The eggs are laid in long, gelatinous strings, from which the developing larvae emerge as free-swimming tadpoles. As soon as the limbs are fully developed and the tail has been absorbed, the small toads leave the water.

On the higher mountain slopes and plateaux one may be fortunate to see a remarkable small dark brown toad with lighter dorsal stripes, the undersurface being a light brownish-grey. This is the striped mountain toad (*Capensibufo rosei*), restricted to certain mountain ranges of the South-western Cape and about 3 cm in length. It is also named Rose's toad after the pioneer herpetologist who discovered it some 60 years ago in a pool below Maclear's Beacon on Table Mountain. Describing his discovery, Dr Rose recorded that a female

Arum lily frog (*Hyperolius horstockii*). (*Photo: A de Villiers*)

Striped mountain toad (*Capensibufo rosei*). (*Photo: A de Villiers*)

mountain toad produced the astonishing number of 7 000 eggs at a spawning. About 6 weeks after the eggs are laid, the tiny toads, about the size of a grain of wheat (6,5 mm), emerge and leave the water to commence their terrestrial life.

Of similar build to toads but usually on a smaller scale is the odd-looking rain frog or blaasop (*Breviceps*). They live underground in loose soil or under stones and are unusual in that they do not breed in water. *Breviceps*, so named on account of the short head, is a small, dumpy frog with short legs. It does not hop but moves like a tortoise. When alarmed, it inflates its body to treble normal size, hence its local name of blaasop. They live on worms, beetles, insect larvae and termites. The eggs are large and encased in jelly and are deposited in an underground nest. The entire development takes place within the egg, the young emerging as fully formed small frogs.

Rain frog (*Breviceps rosei*). (*Photo: John Visser*)

Three species of *Breviceps* occur on the Cape Peninsula, two of which are associated with mountainous habitats, while the Cape rain frog usually occurs on the lower slopes and even in suburban gardens. Unfortunately, these attractive amphibians are steadily decreasing in numbers due to habitat destruction, injury by garden implements and the increasing use of insecticides. The name rain frog is derived from the peculiar call uttered prior to or during rain. Breeding takes place during the wet winter months.

Fish

Visitors to the Cape may be surprised by the dearth of freshwater fish in the mountain streams of the Peninsula. The only fish likely to be seen is the small, minnowlike Cape galaxia. These golden-brown fish with darker brown vertical bars are semi-transparent, the red gills and silvery lining of the body cavity being clearly visible due to the lack of body scales. Galaxias seldom exceed 5 cm in length. They are hardy fish, inhabiting the pools of the peat-stained, acidulous mountain streams and feeding on invertebrates. The eggs are large in proportion to the size of the females, and breeding takes place throughout the summer. Of particular interest is the fact that the genus of *Galaxia* is also found in Australia, New Zealand and the southern tip of South America – further proof that these continents were once part of the ancient land mass of Gondwanaland.

A walker's paradise

JOSE BURMAN

The only way to experience Table Mountain is to walk. To escape the tainted breath of the city, its teeming millions, its noisy traffic, and the fumes of diesel you have only to head for the mountain. Within minutes you can be following a path through thick vegetation, enjoying the sight and scent of flowers, listening to bird-calls or wandering high above the city entranced by the superlative views the mountain provides.

There is mountainous territory all the way down the Cape Peninsula as far as Cape Point, but to most walkers Table Mountain comprises the territory between the Table that rears high above the city and the Hout Bay valley, which cuts through the mountain to Constantia Nek. Both Devil's Peak and Lion's Head are connected to this massif and can be regarded as part of it.

A hundred years ago there were scarcely a score of known routes on Table Mountain; today estimates vary, running up to 1 000 routes. Well over 90 per cent are rock-climbing routes, of degrees of difficulty that range up to some of the toughest climbs in the world. For the walker's purposes, however, a very limited number of routes will suffice to give him or her access to the entire mountain, and these we will now discuss. With them at your disposal, Table Mountain is yours to take.

The walks on Table Mountain are divided into three types: the traverses which cover three sides of the massif; the ascents; and walks on the summit.

The traverses

THE PIPE TRACK

Probably the best-known walk on Table Mountain, the Pipe Track is just that – a track laid in the 1880s, along the route of the pipeline that brought Table Mountain water from the Woodhead Tunnel to the city.

The Pipe Track leaves Kloof Nek opposite the bus terminus, starting with a set of steps ascending past the houses. The track levels off opposite the Mocke Reservoir, and commences contouring the slopes of the mountain. The pipeline crosses several little ravines on aqueducts, with the Pipe Track following faithfully alongside. Soon you reach the Kloof Nek filtration plant. Built in 1938, this plant treated the brown water from the Woodhead Reservoir, with the anomalous result that whereas Cape Town's main supply from Steenbras was still brown, the water from Table Mountain was crystal clear.

Next the path drops steeply into Diepsloot to ascend the slope on the other side, passing under Cairn, Fountain and Grotto Buttresses. Peeping through a gap you

The Pipe Track and the pipeline carrying Table Mountain water to the city. (*Photo: Jose Burman*)

will be able to see the upper cableway station. The path winds on, executing a hairpin bend, while the pipe crosses another aqueduct. The trees thin out and the slope is thick with protea bushes and a multitude of flowers – ericas, brunias, pelargoniums, and at some seasons watsonias and bobbejaantjies.

Now the mountain retreats to form a large kloof – Blinkwater Ravine. Beyond Blinkwater stretch the buttresses of the Twelve Apostles – Porcupine, Jubilee, Barrier, Valken, Kasteel, Postern, Wood, Spring, Slangolie, Corridor, Grootkop and Separation, ending at Llandudno Corner.

The Pipe Track passes below Porcupine, Jubilee, Barrier and Valken Buttresses. Just below Barrier a path branches off to the left: this is the Kasteelspoort route, the easiest and most popular of the Twelve Apostles ascents, and we will be following it later.

Not far beyond this turn-off the Pipe Track debouches into a broad road coming from the right – a relic of the construction of the Apostles Tunnel. For 500 m you follow this road to the point where stone terracing on the slope above indicates the exit of the Apostles Tunnel. Beyond the next corner is a large round-domed stone building standing in the path. This is a break-pressure tank built at the time the pipeline was constructed.

Shaded by tall trees and bushes the Pipe Track winds round the corner to the foot of Woody Ravine. Another aqueduct crosses this ravine, after which the track narrows and steepens as it traverses Spring Buttress. The view is outstanding and varied – Bakoven far below you, with Lion's Head coming into sight as you rise higher, and Robben Island peeping over its shoulder.

A last bend, and you are in Slangolie Ravine. This

The Pipe Track and Kasteelspoort

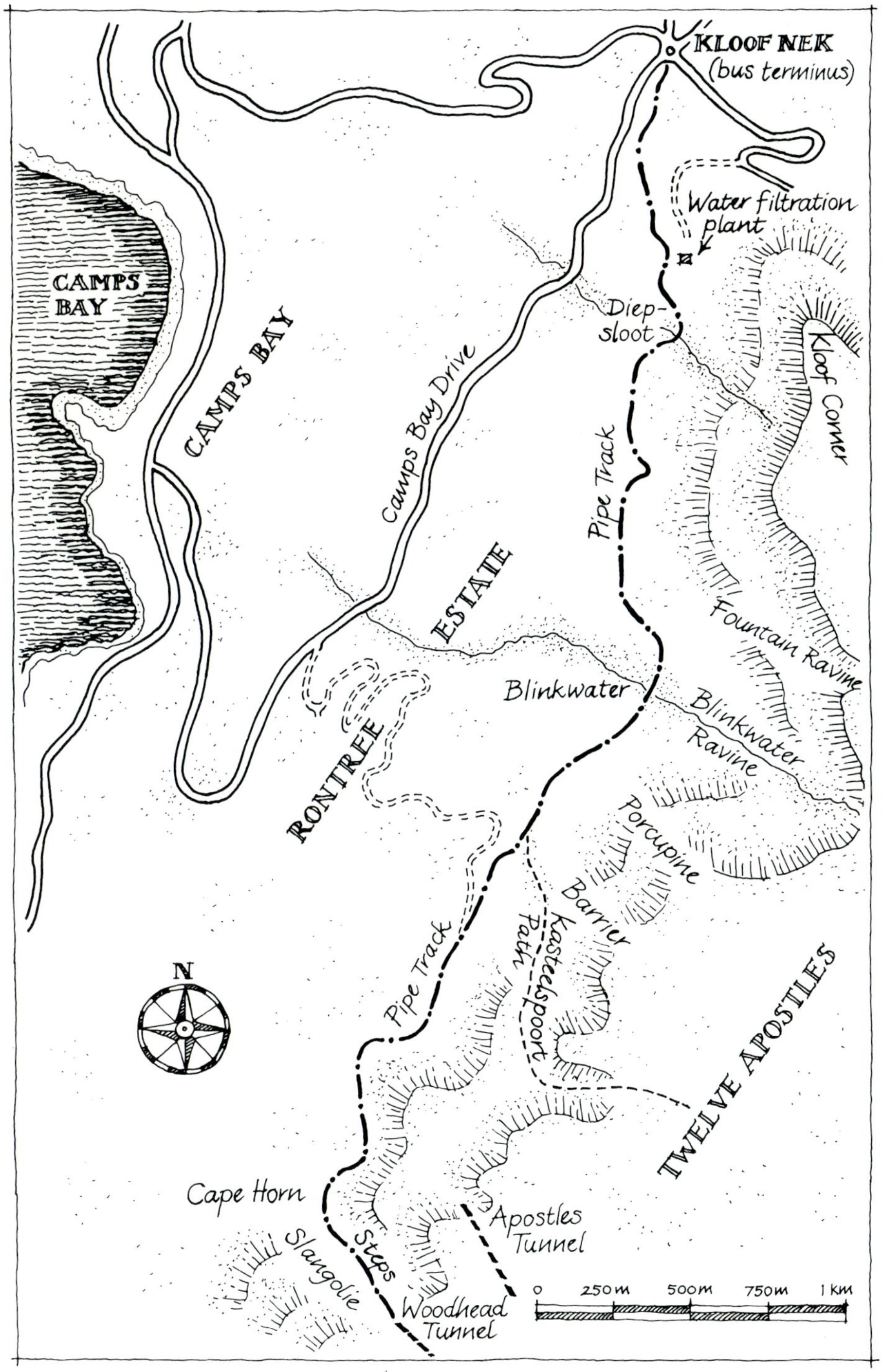

The Twelve Apostles and Camps Bay as seen from the Pipe Track. (*Photo: Mike Scott*)

corner was called Cape Horn by the track-layers, and if you come round it during a south-easter you will agree with their judgment. Slangolie marks the end of the Pipe Track, for this is where the Woodhead Tunnel begins. But before you reach the entrance of the tunnel you will have to brave 540 steps, of which all but 21 lead upwards – stone steps, sloping wooden ones, and finally iron rungs. The whole kloof is a little hair-raising, the path having been swept away by landslides in the past.

The distance from Kloof Nek to the Woodhead Tunnel is under 7 km, and you now face a return along the same route, but with different views as you are facing in the opposite direction.

In 1888 the *Cape Argus*, referring to the Pipe Track, wrote: "This path, when completed, will no doubt be recognised as the promenade of the inhabitants of the suburbs of the future." Looking back from our vantage point a century in the future, we can vouch for the accuracy of this forecast.

THE UPPER CONTOUR PATH

Well above Tafelberg Road there is a path traversing the face of Table Mountain, closely hugging the cliffs – the Upper Contour Path.

Tafelberg Road leaves Kloof Corner, executing a double hairpin bend to reach the lower cable station. At

Start of the Upper Contour Path

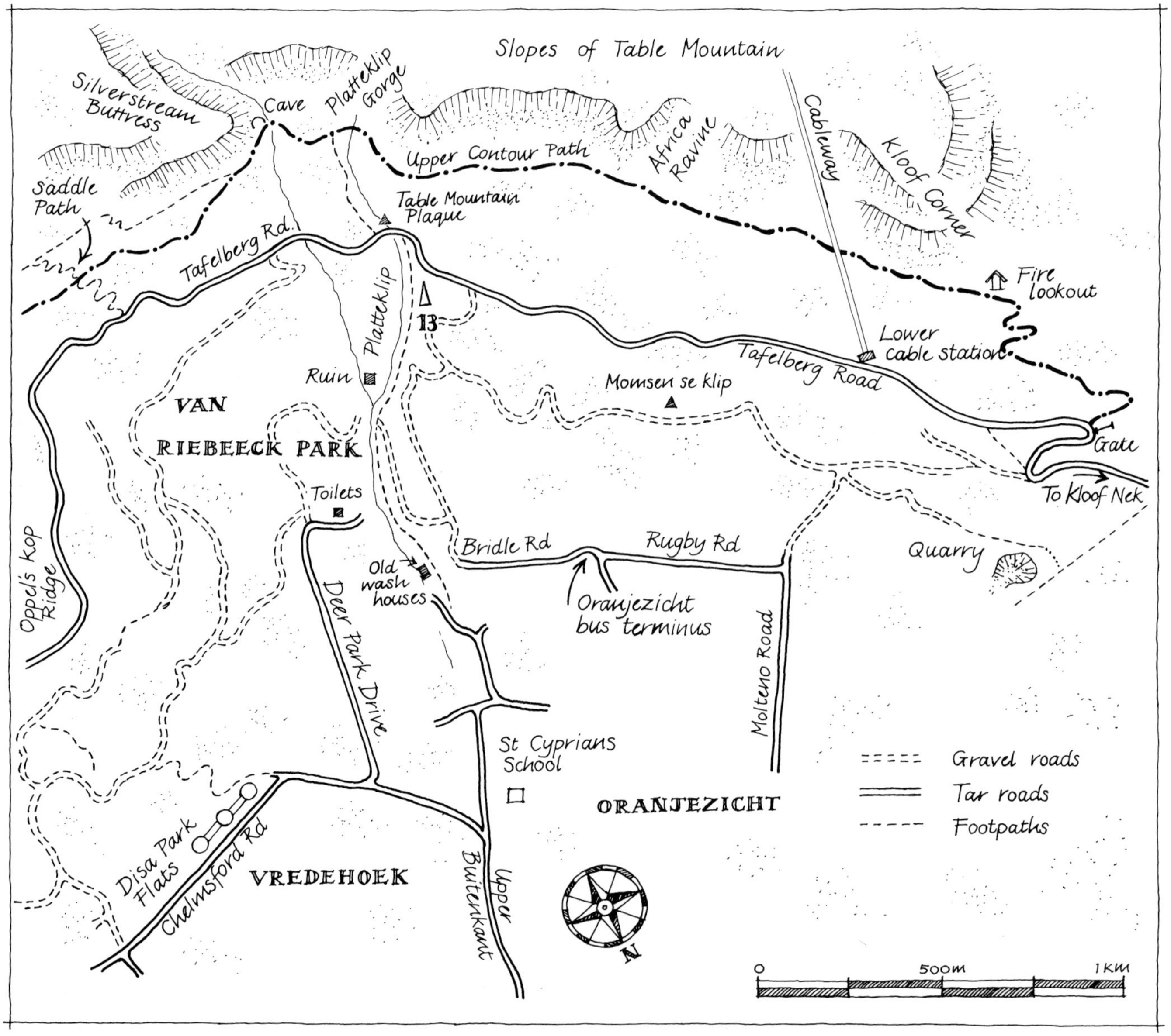

Lion's Head as seen from Kloof Corner. (*Photo: Jose Burman*)

From Silverstream Cave overlooking the city. (*Photo: Jose Burman*)

the second bend of the Z is a flight of worn cement steps climbing the slope from the roadway. The path rises steadily up the slope in a series of wide zigzags, the surface nicely consolidated with wire and logs. This path leads you to the bottom of a series of 213 stone steps, which end at the base of the cliffs. Here stands a fire lookout hut and this is the true start of the Upper Contour Path.

The Upper Contour Path follows the base of the cliffs closely, passing under the wires of the cableway, allowing you a good view of the cable-cars as they pass overhead on their 5-minute journey to the summit. The path crosses India Ravine, and after ascending slightly, skirts a buttress and crosses Africa Ravine. These ravines are so named because of the resemblance to the shapes of India and Africa which their outlines disclose when seen from the city centre stretched out for your scrutiny far below.

The track continues across the face of the mountain, following the spurs and ravines to reach Platteklip Gorge. Up Platteklip runs the original route of ascent, still the easiest way to reach the summit direct from the city; we will be following this route later (page 116). Beyond Platteklip the Upper Contour skirts the slopes before entering Silverstream Ravine. It passes the Silverstream Cave just below the rather formidable waterfall.

On the further side of the cave the path divides, the right-hand route continuing on the same level, eventually joining the Saddle Path to Devil's Peak; the left-hand route (which you must follow), drops steadily down the slope, crossing the Saddle Path onto Devil's Peak. Far below are the controversial triple towers of Disa Park. The path runs closely beneath the cliffs, and the buttresses bend back disclosing Oppel's Kop and Blockhouse Ridge, separated by an enormous cirque.

Your track continues descending round Oppel's Kop; where the path forks take the left-hand one which arrives at the base of a red cliff, in which you catch glimpses of a cave-mouth (Twelve O'clock Hole). The path ends on Tafelberg Road, a few hundred metres from the terminus parking area.

The trek back along Tafelberg Road to your car is quite a formidable one, so unless you are a really energetic walker, try and have a car waiting for you at the terminus – preferably with a bottle of cold beer!

Rest of the Upper Contour Path and Devil's Peak

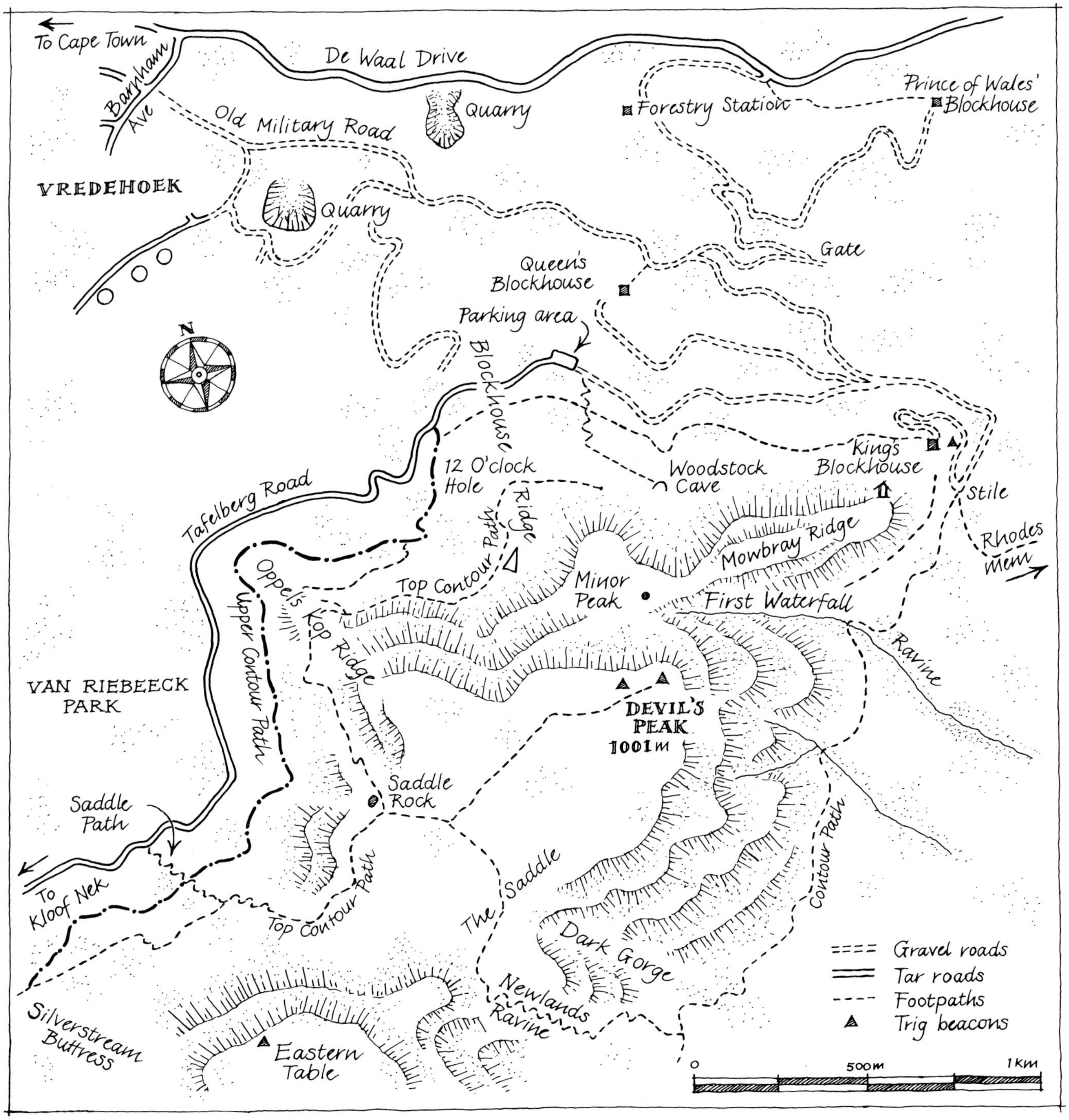

THE CONTOUR PATH

In theory the Contour Path is the same as the Pipe Track – a path running along the slopes of Table Mountain; yet two more different walks it would be hard to imagine. Instead of the open slopes of the Twelve Apostles across which the Pipe Track runs, the Contour Path is a trip through the forest.

The Contour Path begins where the Upper Contour terminates – at the far end of Tafelberg Road, where you can park your car. The view here is particularly impressive, embracing the Atlantic coastline, the harbour and the city centre. The end of the parking area is fenced, and there is a stile over it, which you must cross; soon the road joins that to the King's Blockhouse from De Waal Drive. Not far along this road is a hairpin bend.

The Contour Path

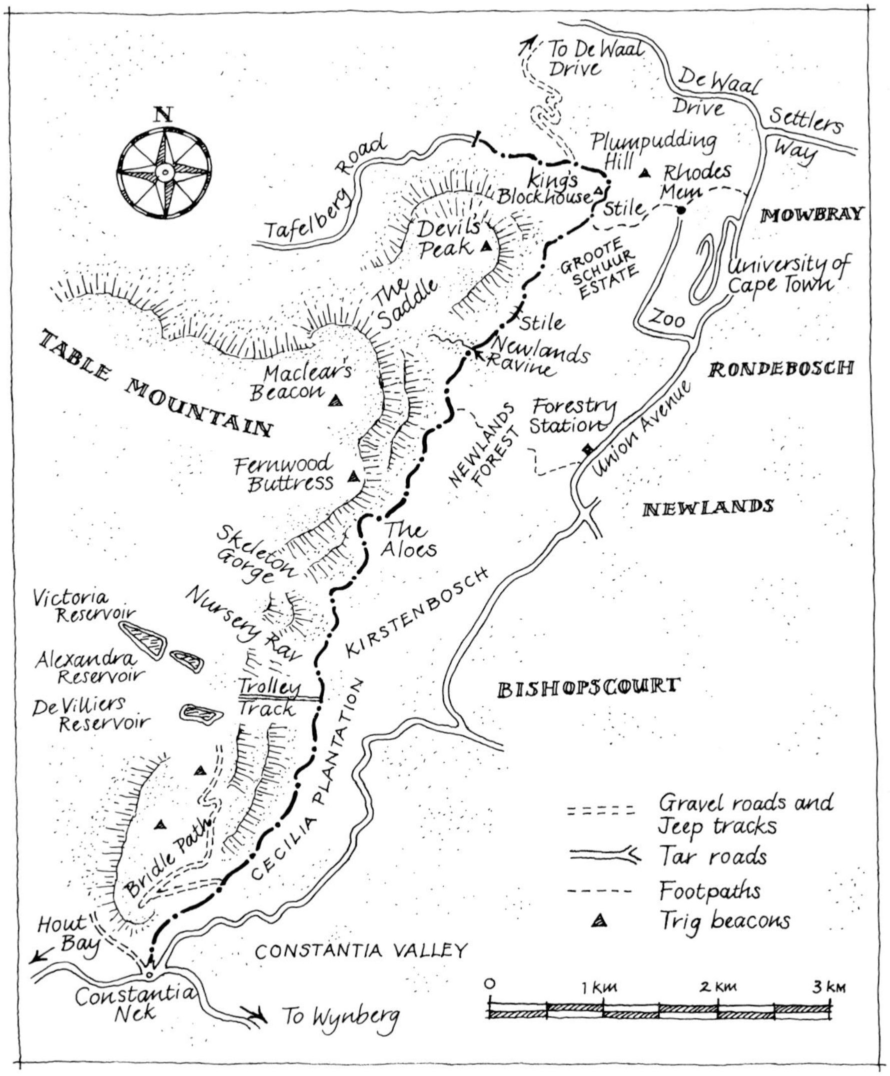

Instead of taking the bend continue straight ahead, crossing a stile that spans the fence.

You are now inside Groote Schuur, the estate that Cecil Rhodes bequeathed to the nation; but for his forethought this lovely mountainside would have been covered with suburban houses. The path runs close under the cliffs of Devil's Peak, descending towards First Waterfall Ravine, to climb the opposite slope and continue on its way. When you cross the next fence you are out of the Groote Schuur Estate.

The trees thin out and you cross Second Waterfall Ravine. The forest here is rich in indigenous trees – keurboom, klipdoring, rooi-els, yellowwood and stinkwood. Below Newlands Ravine the Contour Path crosses a wide band of fallen rocks strewn over the hillside. Across Els Ravine, and past the footpath down through the Newlands forest goes your path. The scene is enchanting, sunlight filtering through the branches, lighting the soft carpet of oak leaves.

The track winds on, across Ascension Ravine, Hiddingh and Wormhole, rising to the 400 m level under the frowning cliffs of Fernwood Buttress. Down to the 300 m level, and across Window Gorge, with the whole sweep of the mountains from Hangklip to the Groot Winterhoek in sight.

The Contour Path crosses Skeleton Gorge in thick forest; below you lies Kirstenbosch, and even these slopes belong to it. Out into the open and the path leads across Nursery Ravine. This is probably the easiest ascent route on the eastern side of the mountain and we will return to it later (page 117). Now the Contour Path drops to a slightly lower level. At the next intersection take the right-hand route, as the other leads to the Forestry Station. The road passes through Cecilia Plantation – a pine forest named after Cecil Rhodes.

You reach the Bridle Path, which ascends the mountain from Constantia Nek. Turn left down this road, which will bring you out at Constantia Nek. The distance is approximately 13 km, so you do need transport to return to Tafelberg Road and pick up your car.

The Contour Path.
(Photo: Jose Burman)

The ascents

THE BRIDLE PATH

The Bridle Path starts at Constantia Nek, where the Contour Path ends. It appears to have originated in the 1890s when the Forestry Department established a nursery of trees on the mountain (above Nursery Ravine).

Opposite Constantia Nek tearoom is a tarred road heading upwards through the trees. After a couple of hundred metres you pass through a gate. Beyond this gate the road continues through the trees of Cecilia Plantation. About 500 m beyond the gate the road forks; take the left branch, and again bear left at the next intersection. Now you are on the regular Bridle Path, which winds back and forth through the trees, till it rises above them and heads north along the mountain slope.

Today the Bridle Path, which was once a packhorse trail, is a wide road, barred to the public as a motor-road but used regularly by the vehicles of the municipality to carry supplies – and fire-fighters when the occasion arises.

The Bridle Path winds along the slope in wide sweeps, rising ever higher, until after a Z-bend it emerges onto the Back Table. On the last section of the ascent the road is concreted, and a substantial bridge crosses a gully in which there is a watercourse.

The Back Table is a level plateau about 300 m below the summit. The Bridle Path crosses the Back Table and, indeed, ends on it. Just before turning onto the main plateau you will see a sheet of water on the left, with a house on the right. The water belongs to the De Villiers Reservoir and the house was built for the waterworks supervisor.

Follow the road across the plateau and you will pass two more reservoirs on the right: the first is Alexandra, and the upper one is Victoria. Beyond the Victoria Reservoir the track turns right to cross a rise and then continues across the wall of the Woodhead Reservoir. This is an interesting trip, with the level waters of the reservoir on the right and the trees of the gorge nearly 50 m below on the left.

Once across the wall, continue up the path which fol-

The Bridle Path

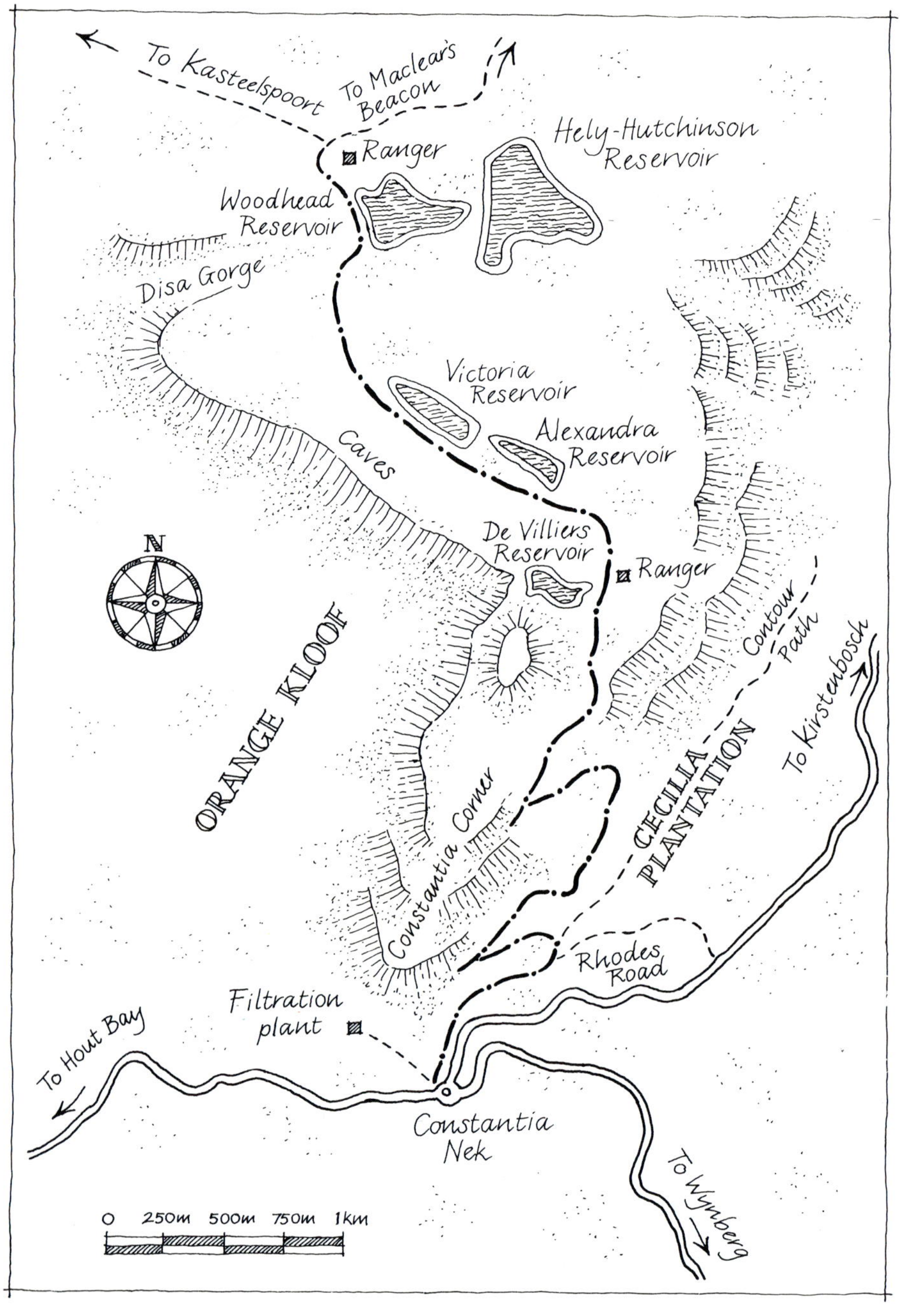

Top right: Alexandra Reservoir in the mist. (*Photo: Jose Burman*)

lows the reservoir fence. About 500 m farther you will come to the gate of the waterworks overseer. Here you join the path from Kasteelspoort and from Maclear's Beacon, which we shall deal with later (page 119).

The distance from Constantia Nek is approximately 8 km, and this is the easiest route up the mountain.

KASTEELSPOORT

The Kasteelspoort path leaves the Pipe Track about a kilometre beyond Blinkwater, and is marked by a sign-post. It is a popular route and the easiest ascent of the Apostles.

Weaving across the slope the path climbs gradually, passing under Barrier, Valken and Kasteel Buttresses. Below lies Bakoven, the most southerly of the Atlantic suburbs, whilst clinging to the shore-line, like a long black snake, is Victoria Road, the last mountain pass of that great pass-builder Thomas Bain.

The path turns up Kasteelspoort, keeping to the left side of the ravine. On a little knoll, halfway up, is a flat platform. Until a few years ago a large post, bearing a rusty wheel, stood here – a relic of the Kasteelspoort cableway.

The walk up the ravine is only a kilometre in length, and is made under the awe-inspiring cliff of Postern Wall, which forms the right-hand side of the ravine. Near the top of the gorge it is necessary to cross the stream to reach the top of the ravine – just where Postern Buttress begins.

At this point it is worthwhile walking a few metres to the right along the top of Postern Buttress, where you will come on the remains of the Kasteelspoort cableway, crouching over the huge drop into the ravine. From here

People standing on the rocky platform near the old Kasteelspoort cableway used during the construction of reservoirs. (*Photo: Mike Scott*)

Lion's Head and slopes of the Apostles from Pimple Gorge. (*Photo: Jose Burman*)

your way lies straight across the lower plateau, following the raised path that leads from the old cableway. This was the track of a little railway line that used to carry equipment from the cableway to the reservoir while it was under construction. The engine was borne up the mountain piecemeal. Not far from the old cableway station you will cross a little path which runs at right angles to Kasteelspoort. This is the Apostles Path to Grootkop, which links all the climbs and routes on this side of the mountain.

At many points on the plateau you will see the foundations of buildings, the remains of the little township that grew on the plateau during the construction of the reservoir. The Mountain Club of SA benefited from this building spree, and its present hut on the mountain is a survivor of that era.

On the far side of the plateau the track leads to the house of the waterworks overseer. Here the path from Constantia Nek joins it and the two pursue their way together to Maclear's Beacon (page 119).

PLATTEKLIP GORGE

The most obvious route up Table Mountain is that huge chasm that cleaves the centre of the Table – Platteklip Gorge. Starting with Antonio de Saldanha in 1503 all the early climbers used this route.

They were not as fortunate as you, however, as you do not have to start walking at the Castle; instead, you can drive up to Kloof Nek and then follow Tafelberg Road, past the lower cable station. About 1,5 km beyond that station there is a parking place near a concreted pool. This is Platteklip Gorge and the path starts on the left side of the pool, following the stream bed. You will climb a number of rocky steps that have been in use for generations. You will also find here the last water in the ravine.

About 12 minutes walk from Tafelberg Road you should reach the Upper Contour Path. Follow it out of the ravine to the left and you will find the continuation of the Platteklip Path some 50 m farther, more steps leading up the slope to the right.

The path crosses the ravine bed again, below a rock cliff, and then slants diagonally up to the right, following the foot of the cliff until it reaches the wide upper ravine. This upper ravine was described in 1741 by Mentzel as "filled with stones of all sizes, amongst which stunted bushes are to be seen in places. An ascent through the gorge is a sore trial for one's feet. At one place many large rocks have to be climbed, at another one has to make one's way over sharp and flint-like stones."

The stones are still there, but the rocks have now been

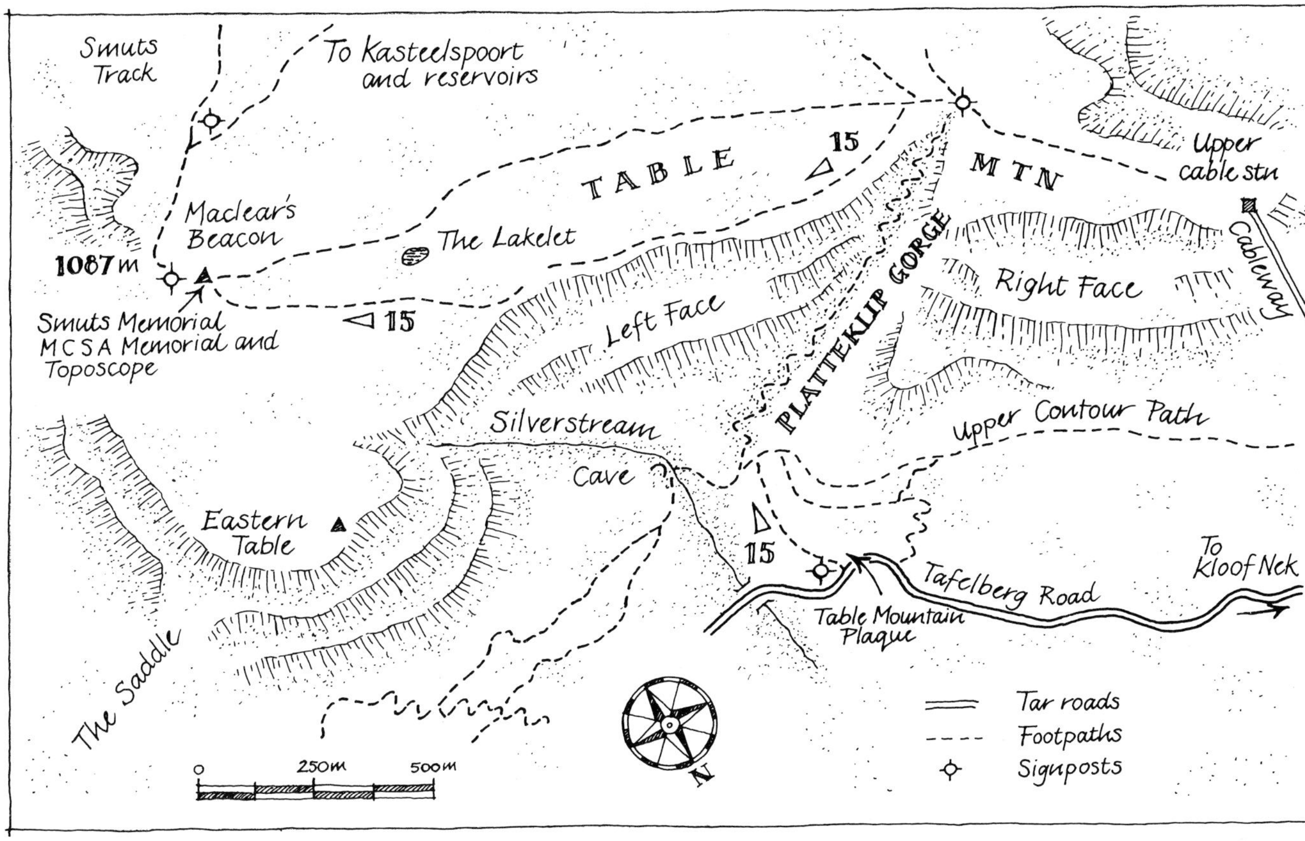

laid into a stone staircase up which you can clamber rapidly though arduously. All of which indicates that while Platteklip Gorge is a simple route of ascent, it is also a fatiguing one – certainly not my favourite climb. Nevertheless it is an easy route of descent, invaluable in bad weather, since you are confined by the walls of the ravine and cannot go wrong.

As you rise higher the gorge steepens, and the cliffs close in to form the "poort" – a narrow gateway onto the summit plateau. There are some 1 800 steps to be climbed, so you may feel that you have earned the fantastic view that awaits you on the plateau.

The gorge debouches onto that plateau about 10 minutes' walk from the upper cable station. The distance from Tafelberg Road to this point is under 3 km – but that is misleading, for you rise some 800 m in that distance.

NURSERY RAVINE

From Kirstenbosch Gardens you can look up and see a buttress which appears to be crowned with the tower-effect of a medieval castle. For years this was known as Castle Buttress but in the 1890s, when the work of reafforestation was at its height, the authorities planted a nursery of trees beside Castle Buttress to test the effect of altitude on different trees. The buttress soon became Nursery Buttress, and the ravine leading to it, Nursery Ravine.

Nursery Ravine starts in Kirstenbosch Gardens, and is the route I recommend for ascending the mountain from the Southern Suburbs. The easiest way is to start, not from the main gates, but from the entrance below the director's house, which stands on a hill south of the main gardens. The road leads up the hill through the gardens, passing the director's house, after which the tar becomes a brick path headed straight at the mountain. Turn left at the T-junction for about 20 m, then right up a dirt road for 100 m. Here is another junction where you turn left; about 50 m along this road you will see a couple of stone steps leading to a path running diagonally up the slope. This is the path to Nursery Ravine.

The path runs at a steadily rising angle and joins the Contour Path, which you follow to the right for about 100 m, when you enter a heavily wooded gorge – Nursery Ravine. Cross the rivulet which flows down the centre, to a flight of stone steps. Start climbing, but do

Nursery Ravine

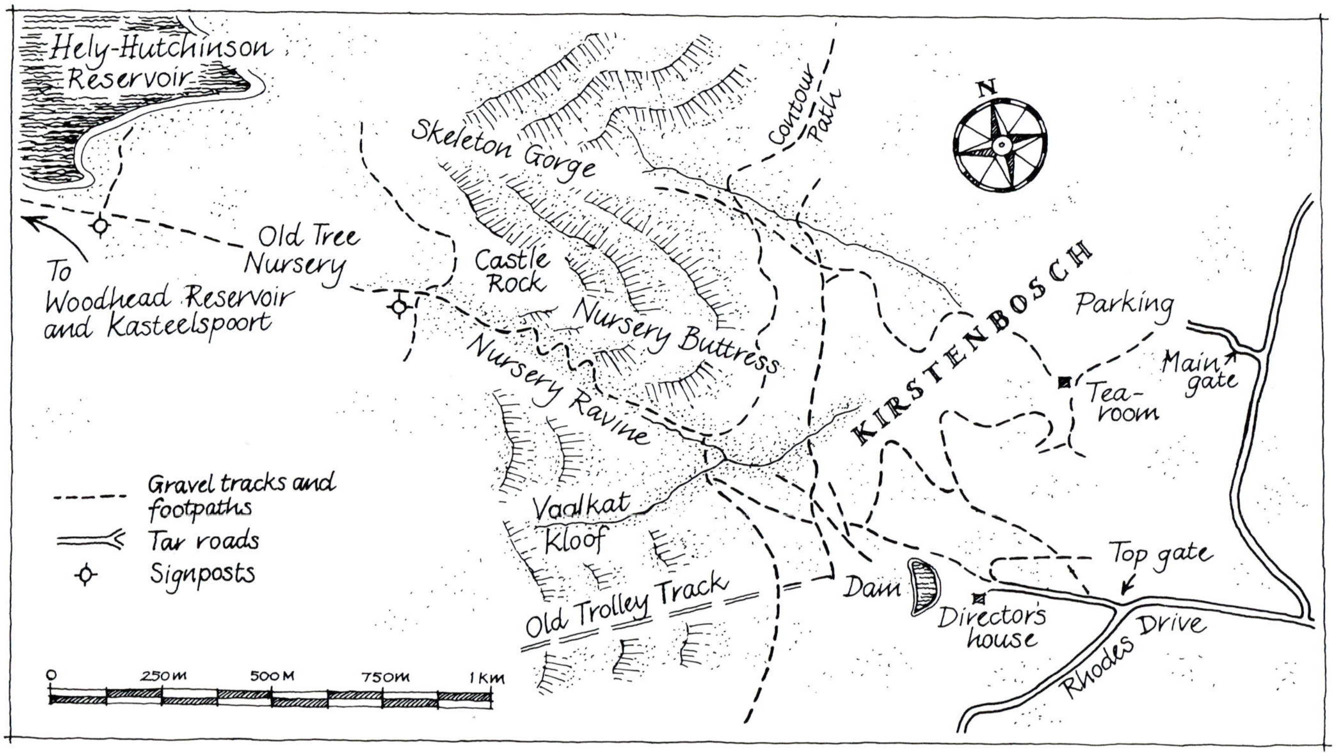

Woodhead (back) and Hely-Hutchinson Reservoirs from Reserve Peak. (*Photo: Mike Scott*)

not bother to count steps – there are over 800! Instead, concentrate on the natural beauty: the rapids, the little waterfalls, the quiet pools in this pocket of indigenous vegetation where wit-els, boekenhout, assegai, wild olive and yellowwood flourish.

Eventually the path leaves the shelter of the forest and crosses the open slope high up the ravine in two long zigzags; the view is breathtaking – all the way across the Cape Flats to the Hottentots Holland and False Bay.

You reach the foot of a stone scree, which has to be climbed; then the path crosses the ravine, which soon widens out. Tall trees are visible ahead as well as broken walls, for this is the site of the one-time nursery. The path narrows again and you cross a nek. A few hundred metres brings you to a fence behind which you can see the waters of a reservoir – the Hely-Hutchinson. There is also a signpost indicating Skeleton Gorge and Maclear's Beacon to the right, as well as Kasteelspoort ahead.

Continue straight, keeping to the left of the fence, and you will soon pass the wall of the Hely-Hutchinson Reservoir. In a little while you will see the wall of the Woodhead Reservoir ahead and come to the Bridle Path where it reaches the Woodhead wall.

The time from Kirstenbosch should be from 90 minutes to 2 hours, depending on the speed at which you travel.

The summit walks

Walks on the summit plateau can obviously be reached from the various ascents already described, but for many people this takes too long and is too strenuous, so the summit walks would be barred to them unless there were another way to do them. Fortunately there is, for you can use the cableway to ascend (and descend) the mountain, thus allowing you the freedom of the lovely walks on the summit plateau. So we will start these walks from the upper cable station.

MACLEAR'S BEACON

The highest point on Table Mountain (1 087 m), Maclear's Beacon is a "must" for all Table Mountain walkers. It was set up in 1844 by Sir Thomas Maclear, Astronomer Royal at the Cape, and is about 45 minutes walk from the upper cable station.

As you leave the cable station the restaurant lies straight ahead, while on the left is a flight of stone steps which climbs between the telescope platforms and the restaurant to emerge on the flat plateau of the western table. The path forks, and you can take either: the right-hand path curves slightly and drops in an easy scramble down a couple of rock ledges to the mouth of a ravine. This is the top of Platteklip Gorge (page 116), some 10 minutes' walk from the cable station.

Maclear's Beacon, set up in 1844 by Sir Thomas Maclear, Astronomer Royal at the Cape, is about 45 minutes' walk from the upper cable station. (*Photo: South African Library*)

Table Mountain's frontal table overlooking Devil's Peak. (*Photo: Jose Burman*)

Straight ahead you will see your path ascending between the bushes out of the nek. Follow it and you will reach the central table of the plateau. The path forks at a signpost – take the left-hand route. Soon you cross a broad track which leads to the frontal table, which you will see on your return. The going is easy, the rocks often covered by springy tufts of grass, the path threading a way between shrubs and flowers.

Ahead and slightly to the left you will see a flat-topped outcrop which rises above the plateau near the eastern end, and is crowned by a pointed pyramid; that is Maclear's Beacon. The path leads directly to the outcrop and up its southern side to reach the beacon itself. The distance from Platteklip is about 2 km.

Maclear's Beacon rises just above an amphitheatre in the rocks where an annual service is held. This used to be attended regularly by Field Marshal Jan Smuts, and a plaque is set up here to his memory. From Maclear's Beacon you get a clear view of the mountains to the east, but if you wish to see the central city you must descend below the Smuts Memorial, where you will find a path angling across the plateau on the northern side of the outcrop. Ten minutes' walk along it brings you to the front table, overlooking the city centre.

The path follows the front table all the way back to Platteklip Gorge, affording some spectacular views, to descend sharply into the nek at the head of the gorge. The entire circuit can by done comfortably in under 2 hours.

MACLEAR'S – SMUTS TRACK – WOODHEAD

You may recall that three of the ascents brought you onto the Back Table. The following two walks are connecting routes between the Back Table and Maclear's Beacon.

When you reach Maclear's Beacon from the upper cable station, you might like to explore the summit more fully, rather than returning immediately to the cable station. In that event continue across the summit outcrop to a signpost and follow the path marked Skeleton Gorge/Blinkwater. This is the Smuts Track, which the Field Marshal was wont to use on his walks to the summit. About 15 minutes down the Smuts Track you will pass a sign to Echo Valley (the next walk).

Continuing down the Smuts Track – quite rough going – you will find yourself sliding down rocks and picking your way among boulders, whilst still admiring the extensive view. The steadily descending path brings you to a wide rocky ledge overlooking a little valley; turn sharp left and the path will take you down to the base of

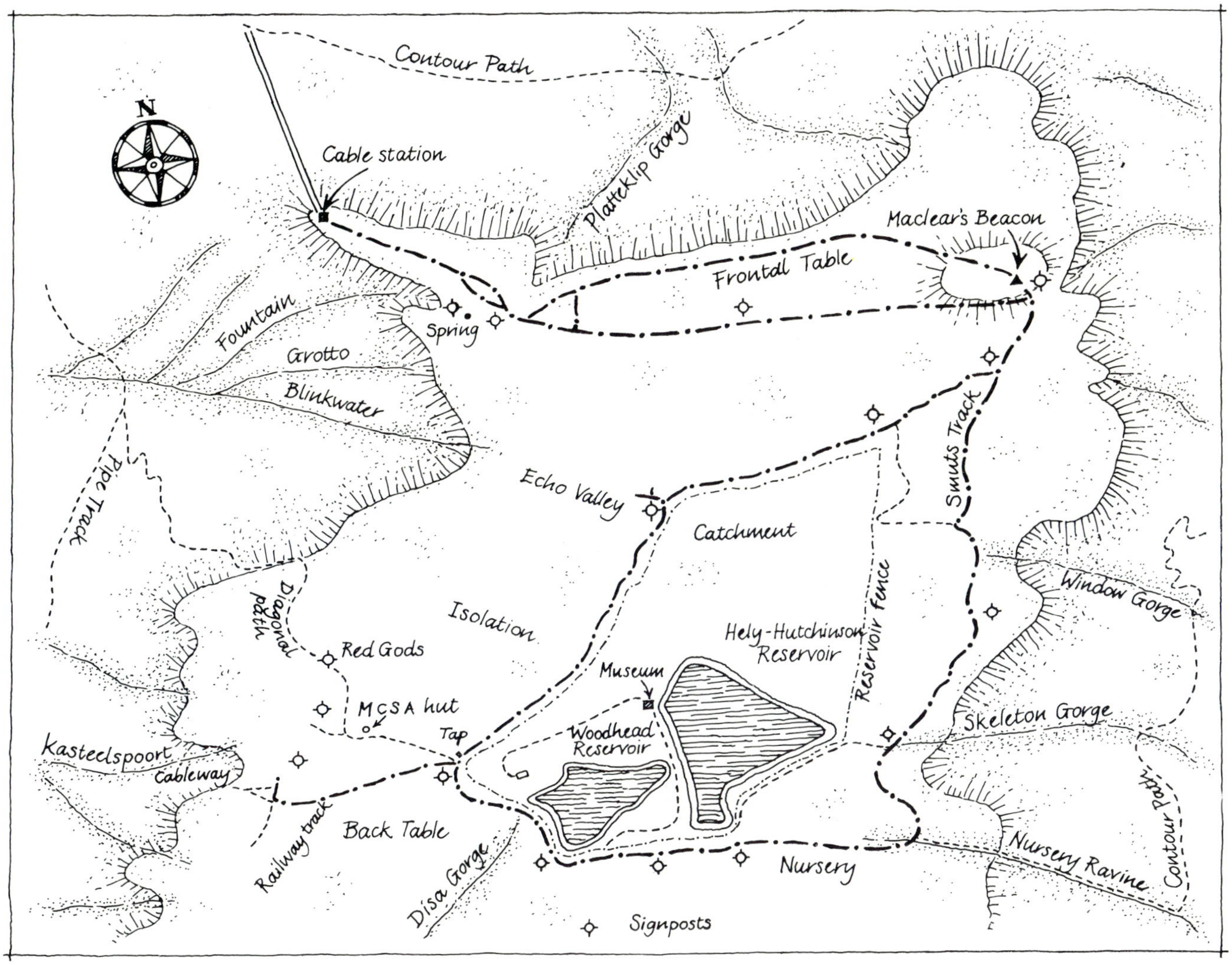

the cliff – an easy scramble. From here a short walk brings you to the stream in the valley. This is Window Gorge (40 minutes from Maclear's Beacon).

The path climbs the opposite side of the little valley and turns half-left. You will pass another signpost marked Skeleton Gorge, as you continue to skirt Junction Peak. Finally you will descend into the ravine which proves to be Skeleton Gorge. As Smuts usually climbed Skeleton Gorge, the Smuts Track turns left at this point.

Slightly below the signpost in Skeleton Gorge is a huge rock mass on the far side of the path. Descend a few paces and skirt round the left of this rock mass and you will find the path again. It climbs the ridge diagonally to the left, crosses the buttress and descends into Nursery Ravine. In the ravine turn right and follow the path (which you have already used in the ascent of Nursery

Top of Window Buttress looking down Window Gorge. (*Photo: Mike Scott*)

Skeleton Gorge Path. *(Photo: Mike Scott)*

Orange Kloof waterworks in Disa Gorge. *(Photo: Mike Scott)*

Ravine), and it will bring you to the wall of the Woodhead Reservoir where the Bridle Path reaches it.

All you have to do now is cross the wall and make your way to the house of the waterworks overseer – the central point on the plateau where the various paths meet.

Time from Maclear's Beacon to this central point is about 75 minutes.

WOODHEAD – ECHO VALLEY – MACLEAR'S

The second route from the Back Table is via Echo Valley, and as you reached the overseer's cottage at the end of the last walk, you can do this Echo Valley route from below, as the two make a nice circuit.

The path follows the fence. You will cross a cemented path (this marks the overseer's water supply), and come to a dirt path slanting off at a slight angle. This leads through the trees to the foot of the peak. The path climbs through thick vegetation, zigzags onto a higher level, and passes round a large rock on a built-up concrete platform. A small section of safety-chain hangs down a sloping rock which you must ascend. Another bend onto a higher level, and the path traverses round the hill.

The view below is unusual, as you are looking down on the two reservoirs. Just below the Hely-Hutchinson wall you can see the museum building in which stands the railway engine used in building the dams.

The path now turns into a valley on the left and continues up the slope, zigzagging through interesting vegetation and flowers. From the overseer's cottage to the crest takes just over 30 minutes. Beyond the crest you look down into Echo Valley. At the bottom of the valley is a signpost, where you turn right. In winter this can be a wet route, for you are crossing the main catchment streams that feed the reservoirs, and the path is raised on logs for some distance. The walk will take you through an attractive valley, the path rising slightly most of the way.

You will pass a signpost which points to the right, indicating Skeleton Gorge, while ahead it points to Maclear's Beacon. The path rises steadily following a stony stream bed. Another 15 minutes brings you to the

Smuts Track at the signpost to Echo Valley. Turn left up the track and soon you will be back at Maclear's Beacon.

The time for this route is about 90 minutes.

That marks the end of walks on Table Mountain proper, but as they are generally regarded as part of the massif, the ascents of Lion's Head and Devil's Peak are included below.

LION'S HEAD

The walk up Lion's Head is short (660 m), but as there is boulder-scrambling near the top it is recommended only for the nimble of foot.

If you drive up the Signal Hill road from Kloof Nek for nearly a kilometre you will come to a point where a jeep track leaves the tarred road; this is the start of the Lion's Head route. The road climbs diagonally back towards Kloof Nek, and spirals completely round the

The original chains on Lion's Head. (*Photo: South African Library*)

Lion's Head

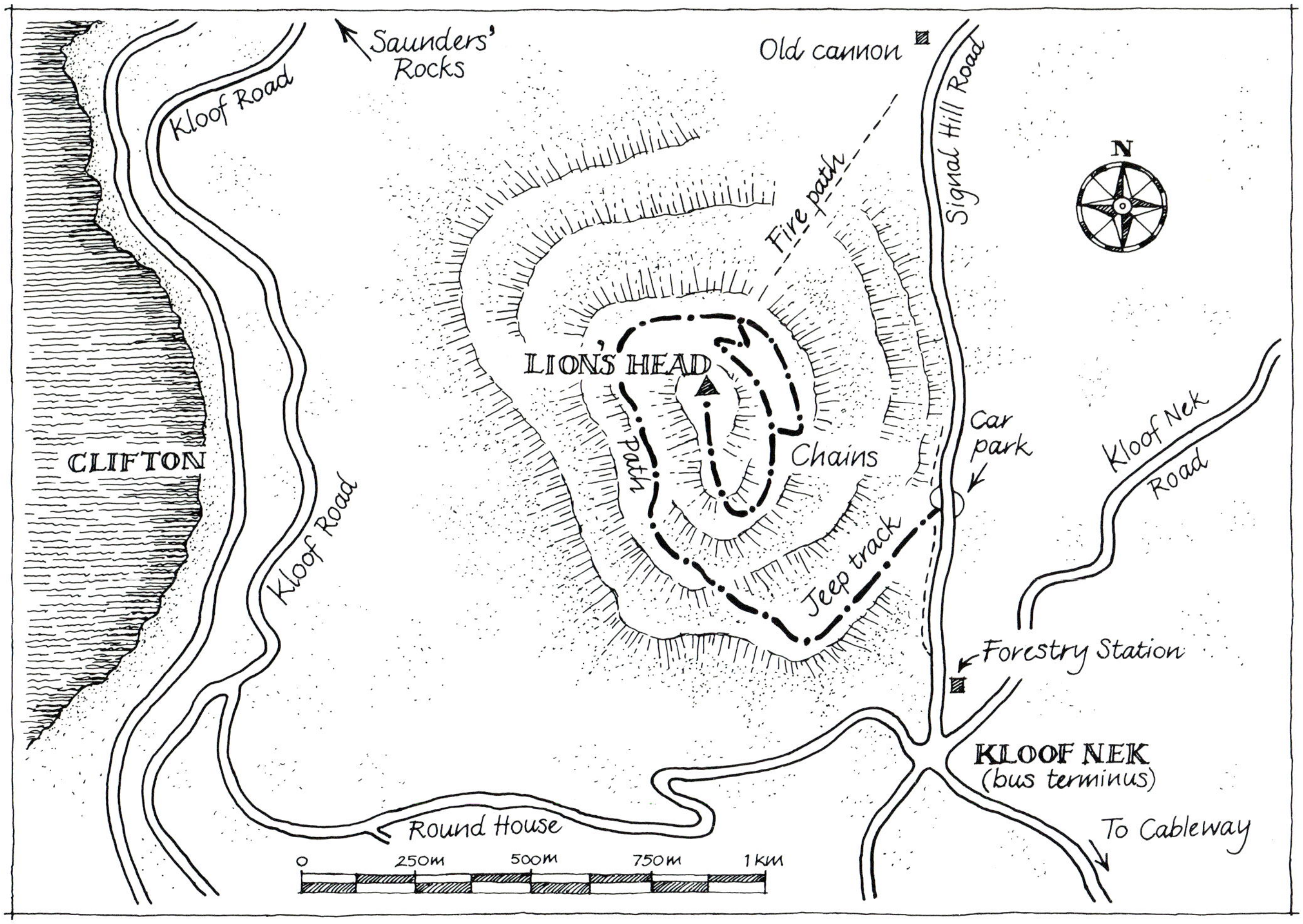

A Kloof Buttress view from Lion's Head. (*Photo: Mike Scott*)

peak, rising steadily. As the going is easy you can admire the view all the way.

Chains were hung on the rocks here in 1881 to assist climbers, but unless you are keen on climbing chains it is not necessary to do so, as both sets of chains can be bypassed. The bypass comes just before a large pine tree at the head of the fire-break leading down to Signal Hill. The path slants diagonally upwards, first left, then right, followed by a short zigzag from which it emerges on a ledge running above the lower cliffs. This bypasses the first set of chains. Continue along the path for a further 300 m, round the corner, and immediately above Glen Beach; here the path turns upwards, an easy scramble bringing you onto the level plateau below the final pyramid and above the second set of chains.

From here the route lies up the corner – a simple matter of clambering up the piled rocks. Whenever in doubt keep to the left of the skyline. At one point there is a short length of chain to assist you up a metre-high pitch. Although the climbing is easy, be careful, as you are on a knife-edge with considerable drops in places.

The total distance is under 4 km and there are few places in the Peninsula where you can see so much beauty for so little expenditure of energy. For the more adventurous a climb up Lion's Head on a summer evening, to descend by the light of the full moon, is extremely rewarding.

DEVIL'S PEAK

Drive along the Tafelberg Road, past the lower cable station, and a few bends beyond Platteklip Gorge you will reach the Saddle Path – a flight of steps and a path marked "To Saddle & Oppel's Kop". This is the best route up Devil's Peak.

The Saddle Path zigzags its way up the slope, interspersed with short sections of stone steps, to level out at an altitude of 700 m. The path leads round Silverstream Buttress and runs into the nek between Table Mountain and Devil's Peak to reach Saddle Rock; this large rock has been used by generations of climbers as a rendezvous. Time: about 45 minutes from Tafelberg Road.

From Saddle Rock there are several paths leading up the peak but it is best to choose one that runs diagonally rather than straight up, as the slope is steep and there is still a distance to climb. However, within an hour from

Oppel's Kop on Devil's Peak. (*Photo: Mike Scott*)

Sunset over Devil's Peak. (*Photo: Jose Burman*)

Saddle Rock you should be resting on the summit of Devil's Peak.

Only slightly lower than Maclear's Beacon, Devil's Peak at 1 010 m possesses a commanding view – far more widespread than any point on Table Mountain itself, and the peak is well worth the effort.

Tips on walking

In walking, foot comfort is the primary consideration. Roomy boots or stout, comfortable shoes with rubber soles are essential, and thick socks will save the discomfort of blisters. If you are properly shod, you will never know what blisters are. At the first sign of discomfort, do not be ashamed to stop, remove your boots and socks, and check the trouble. Always carry some adhesive plasters; applying one of these to a spot which is chafing will usually prevent a blister forming.

On Table Mountain carry a light pack, if possible. In this you can carry your jersey (a "must" at all times) and some food. A light plastic mackintosh adds little weight, and gives security against sudden changes in the weather.

A few simple precautions should keep you safe:

- Never walk alone.
- Leave a message saying where you will be walking and how long you expect to be.
- Keep to the path religiously. If you lose the path do not go on, but turn back and find the last bit of recognisable path; then cast around for the continuation. You leave the path at your peril.

Conservation of the mountains of the Cape Peninsula

DR DOUGLAS HEY

The gradual degradation of Table Mountain, as evidenced by the deterioration of the natural plant cover and resultant soil erosion; the disappearance of many species of fauna and flora; devastating veld fires; the rapid spread of invasive alien vegetation; urban encroachment; and acts of wanton vandalism such as the mutilation of trees, defacement of rocks, littering and pollution have been a matter of increasing concern among responsible citizens since the early days. The first "plakkaat" restricting public access to the forests and the exploitation of trees and prescribing severe penalties for firing the veld was promulgated by Commander Jan van Riebeeck as early as 1658.

By 1947 growing public concern regarding the state of the mountain resulted in a meeting convened by the South African Association of Arts attended by delegates from 27 societies. A resolution was passed urging the government to appoint a commission to investigate the preservation of Table Mountain and its lower slopes as a national asset. This led to the appointment in 1951 of

Scenic grandeur of the mountains. *(Photo: Cape Nature Conservation)*

Soil erosion on the slopes of Devil's Peak. (*Photo: Cape Nature Conservation*)

Anti-erosion work on Devil's Peak. (*Photo: Cape Nature Conservation*)

the Van Zyl Commission, which recommended the establishment of the Table Mountain Preservation Board. This body was duly appointed in 1952 by the Minister of Education, Arts and Science. Its functions were largely advisory and administrative, but it also provided a forum for coordinating the activities of the various landowners on the mountain.

Major objectives of the Preservation Board were the protection and consolidation of mountain land above the 500 ft contour and the acquisition of private properties within the 5 516 ha of the proclaimed Table Mountain conservation area. This was made possible by a subsidy formula in terms of which the State, the Cape Provincial Administration and relevant local authorities each contributed a third of the costs. Credit is due to the Board for acquiring a number of properties, enabling the proclamation of the mountain as a national monument in 1957. The Board was also responsible for preventing the erection of unsightly structures on the upper slopes of the mountain. Due to lack of funds, however, little attention could be devoted to the control of erosion or the spread of alien vegetation.

The continued deterioration of the mountain did not pass unnoticed by the authorities. In June 1963 the Administrator of the Cape, Dr Nico Malan, convened a meeting of interested bodies to discuss the problem at the instigation of the provincial secretary, Mr W J B Slater, and the Cape Town City Council. Resulting from this meeting, the Table Mountain Nature Reserve was proclaimed in 1963 with a grant-in-aid for management purposes on a rand for rand basis. This was followed by the proclamation of the Silvermine Nature Reserve in 1965, which extended the protected area.

A decade later, the Cape Town section of the Mountain Club of South Africa commissioned a report on the general state of the Peninsula mountains. In addition to the problems of uncontrolled fires, soil erosion and the spread of invasive alien vegetation, this report drew attention to new threats including overgrazing by Himalayan tahr, the depredations of baboons, erosion caused by trail bikes, and littering. More recent reports on Table Mountain are the excellent contributions by Moll and Campbell (1976) and Moll and McLachlan (1977), the latter dealing in detail with the network of paths on the mountain. These reports were studied by the Table Mountain Preservation Board and the majority of the recommendations were accepted, with a few minor amendments. However, lack of funding prevented their implementation. The Board was disbanded in 1984 and its functions assigned to the management committee for the Cape Peninsula Nature Area.

The necessity of consolidating the conservation of the

Peninsula mountain chain, including Table Mountain, under a single authority in order to promote the implementation of a uniform management policy including the control of veld fires, the eradication of invasive alien vegetation, the prevention of soil erosion, the monitoring of development and control of the public was now generally accepted by the authorities. After consultations between the ministers of the state departments concerned and the Administrator of the Cape, it was decided in 1977 by the Minister for Planning and the Environment to appoint a one-man commission in the person of Dr Douglas Hey to investigate and report on the future management and control of the mountains of the southern Peninsula.

In its report, published in April 1978, the Hey Commission found that the environmental degradation of the mountain was due to increasing human overuse, residential township development and inadequate management. The report strongly urged that recreation areas be considerably enlarged by the proclamation of the entire Peninsula mountain chain above the 152 m contour from Signal Hill to Cape Point as a nature area in terms of Section 4 of the Physical Planning Act of 1967. (Such areas have since been renamed as Protected Natural Environment – PNE – in terms of the Environmental Conservation Act of 1989.) The commission found that, although this presented the ideal solution, it would be financially impracticable to purchase all this land for a nature reserve and place it under a single authority, as had been urged by many of the societies consulted. Not only would this have required a vast sum of money, but it would also have involved protracted and costly legal negotiations.

It was therefore proposed as an interim measure that the proclaimed area be managed on the basis of a federation of all landowners. Landowners would not lose their existing rights and would be permitted to continue their activities, provided that such activities did not conflict with a conservation policy applicable to the entire nature area. Each landowner would be required to implement approved management plans based on this policy and financed by the state.

The commission urged that the conservation of fauna and flora receive prime consideration, that public recreation be secondary, and that no activity or development to the detriment of the natural environment be permitted. It recommended that management be based on a combined programme of nature conservation and outdoor recreation, embodying the principle of multiple usage. The commission further proposed that the existing boards be amalgamated into a single management committee for the Cape Peninsula mountains.

The commission also addressed many other matters relating to the conservation of this unique area, including the control of public access and behaviour; the provision of roads, paths and facilities for camping and picnicking; the eradication of alien fauna and flora and the reintroduction of native species; and the preservation of historic features and structures. Those seeking further details should consult the report.

These recommendations were duly accepted by the government with one exception, viz. that the area be placed under the custodianship of the Minister of Environment Affairs and not the Administrator of the Cape as proposed. The area was proclaimed a nature area in 1983 and a committee advisory to the Minister of Environment Affairs was appointed in terms of the Environmental Conservation Act of 1982. The committee was initially comprised of representatives of all landowners, local authorities and organisations concerned with the conservation and utilisation of the area, under an independent chairman appointed by the minister. The committee was empowered to appoint subcommittees of specialists for specific tasks. The Department of Forestry was nominated to serve as the executive authority. The management of the nature area was subsequently devolved to provincial level and the committee was made advisory to the Administrator of the Cape, with the Chief Directorate of Nature and Environmental Conservation as the executive authority.

Among the first tasks of the management committee was the delineation of the boundaries of the nature area. In formulating its proposals, the committee was directed to use existing cadastral beacons as far as possible to avoid the expense of surveys to establish new beacons, and the delays this would have caused. Consequently, there are sections where the boundary of the nature area is not satisfactory where it cuts through natural systems. As finally approved, the nature area is approximately 30 000 ha in extent, including all land above the 152 m contour from Signal Hill to Cape Point. At places such as Sandy Bay and Cape Point, however, it extends right down to the sea, thus adding to the diversity of the habitat being conserved and augmenting its recreational importance. Although Karbonkelberg is cut off from the mountain chain by urban development, the slopes sweeping down to the sea are still relatively unspoiled, and, together with Sandy Bay and the adjoining land, form an important component of the nature area. Scenically, Sandy Bay is most attractive and should be conserved for recreational purposes.

Another important project was to formulate a general conservation policy for the nature area on which future management plans could be based. In essence conser-

A paved footpath on top of the mountain. (*Photo: City Parks and Forests*)

A consolidated footpath on Lion's Head. (*Photo: City Parks and Forests*)

vation is interpreted as the sound and wise management of the environment, involving not only its protection but also its usage. Once this policy had been formulated and formally approved by the Minister, attention was devoted to preparing management plans for specific conservation projects such as the control of soil erosion, combating veld fires, the eradication of invasive alien plants and the planning and consolidation of a system of footpaths extending from Signal Hill to Cape Point.

The present network of footpaths within the nature area – some of which are badly eroded – is far too extensive and must be reduced. The short cuts taken by some hikers across zig-zag paths and along fire-control tracer belts are also resulting in erosion. A footpath plan for the entire nature area was prepared by a working group of the management committee under the convenorship of Mr Paul Britton of the Department of Parks and Forests of the City Council. All badly eroded and unnecessary paths will be closed off and revegetated. The paths included in the plan will be upgraded and paved, using natural materials as far as possible – a costly project, but essential for the conservation of the nature area. This initial report has recently been updated and enlarged by the recreational subcommittee under the convenorship of Mr Lester Coelen. The entire footpath system has been mapped and will be published in two sections, thus providing the public with the first accurate map of the approved footpaths on the mountain. Users of the mountain will be required to remain on these footpaths. Most young fynbos plants are very brittle and even a single hiker striding across the veld can cause extensive damage. The committee was not in favour of building rest huts within the nature area, and recommended that overnight facilities for hikers should be provided at convenient points outside the nature area by private enterprise, in conformity with government policy on privatisation.

A management plan for the systematic eradication of invasive alien vegetation over a 15-year period is already being implemented by state and semi-state landowners, and private hack groups are also playing a significant part in the programme. Unfortunately, cleared areas are becoming rapidly revegetated by seed banks in the soil and as a result of dwindling financial resources, the plan is lagging badly.

The most dangerous invasive aliens of the Peninsula mountains are the three major species of Australian acacia (*Acacia cyclops*, *A. saligna* and *A. melanoxylon*); the stinkbean (*Albizia*); the Australian myrtle (*Leptospermum*); three species of hakea; and the cluster pine (*Pinus pinaster*). The eradication of established stands of alien vegetation is an arduous and expensive task. In addition

The eradication of alien vegetation. (*Photo: Cape Nature Conservation*)

Spraying alien vegetation with herbicide. (*Photo: Cape Nature Conservation*)

to the labour involved in felling and stacking, the stacks must be burnt once they have dried out. Each year thereafter for at least 5 years there must be follow-up operations of hand weeding and spraying with herbicides. The seed banks in old stands of acacia may produce seedlings for up to 10 years or more. Recently encouraging success in the biological control of invasive aliens has been achieved with beetles which reduce seed production, a fungus (*Colletotrichum*), which kills hakea, and the Australian gall rust (*Uromycladium*), which attacks the Port Jackson willow (*Acacia saligna*). The most effective means of eradicating *Acacia cyclops*, however, is by hand as the plants do not coppice when chopped at soil level. Following on the eradication programmes of the management authorities, the re-establishment of the natural fynbos is encouraged by reseeding cleared areas and the planting of yellowwoods, wild peach (*Kiggelaria*), silver-trees and species of proteas, heaths and bulbous plants in suitable habitats. Thousands of indigenous trees and shrubs have been planted in suitable areas that have been cleared of aliens, with the objective of not only re-establishing the fynbos but reintroducing a number of species, such as the sugarbush (*Protea repens*) and the striking *P. grandiceps*, which were once common on the mountain slopes but have subsequently disappeared.

Australian gall rust (*Uromycladium*). (*Photo: Martin Pollack, City Parks and Forests*)

Great credit is due to the authorities for their work over many years in clearing alien vegetation, replanting indigenous trees, checking soil erosion and consolidating footpaths. One of the first areas to be cleared by the Department of Parks and Forests of the City Council of very dense stands of pines, interspersed with invasive blackwood, was the 200 ha surrounding the old reservoirs on the back plateau of Table Mountain. This operation has been a great success. The fynbos has regenerated remarkably well, even to the extent that species such as the silver pea (*Priestleya tomentosa*), considered to have become extinct on the mountain, have reappeared. The Department of Forestry has cleared the very precipitous south-eastern slopes from Devil's Peak to Kirstenbosch of invading pines. In the Cape Point area the Divisional Council (later the Regional Services Council) has cleared over 5 000 ha of veld of alien acacias. Credit is also due to volunteer hack groups who have rendered sterling service over many years in clearing specific areas of invasive aliens and re-establishing the native flora. An outstanding example is the northern slope of Elsie's Peak above Fish Hoek.

The striking *Protea grandiceps*. (*Photo: Cape Nature Conservation*)

The need to coordinate fire protection measures and fire-fighting has long been recognised, culminating in the establishment of the Cape Peninsula Fire Protection Committee in 1949. The first task of this committee was to prepare fire protection plans for the entire Peninsula mountain range, which have subsequently been updated every five years. Finance to subsidise the work of the local authorities is provided by the state. The provision and upkeep of a network of fire-breaks, fire lookouts and radio contact have enabled fire-fighters to bring many fires under control at an early stage, and the activities of the committee and its member bodies have considerably reduced the number of fires each season.

Well-established fynbos vegetation is highly inflammable and while fires may be started by natural causes such as lightning and rolling boulders, more than 95 per cent of fires result from man's activities. Fires raged in the mountains of the Western Cape long before the first white settlers landed in Table Bay, and were often started by Khoikhoi herdsmen to provide grazing for their cattle. Although fynbos is well adapted to fire, if the veld is burnt at short intervals those plants that require years to mature and produce seed will gradually be eliminated. Such fires also result in a heavy mortality of smaller creatures, nesting birds and reptiles which are unable to escape the flames.

With the increasing use of the Peninsula mountains for recreation, veld fires have become an annual occurrence, caused by careless picnickers and smokers. In addition to strict legislation prohibiting the lighting of fires except in fireplaces at approved picnic sites, and an education programme to promote public cooperation, it is also necessary to manage the vegetation by planned and systematic controlled burning, on a rotation of 15 to 20 years. This not only diminishes the risk of devastating fires but also encourages the widest diversity of flora, thus providing a better habitat for wildlife. The coordination of the control of veld fires on the Peninsula mountains remains the responsibility of the Cape Peninsula Fire Control Committee.

From the outset, the management committee has been deeply involved in considering proposals for development both within the nature area and on the perimeter. While the rights of the individual landowners within the Cape Peninsula Protected Natural Environment (CPPNE) are recognised and must be honoured,

A devastating veld fire. (*Photo: Cape Nature Conservation*)

no development which could adversely affect natural systems, the habitat of rare fauna and flora, scenic or historic features or result in the general degradation of the environment can be supported. Any development within the nature area must blend with the environment and be aesthetically acceptable. Natural materials must be used as far as possible and, after completion of all works, the landscape must be restored. Within the nature area there are some sensitive ecosystems where no development of any kind can be considered. There are a number of species of plants and animals that are rare and restricted in habitat which could be adversely affected by development. For example, the ghost frog (*Heleophryne rosei*) is found only in certain ravines on Table Mountain. A proper survey of the fauna and flora of the Peninsula mountains is therefore essential as a basis for preparing sound management plans. Such areas should be acquired by the state, and the landowner should be compensated. Developments on the perimeter of the nature area which might adversely affect the scenic beauty are also considered, and every effort is made to ameliorate the impact of such developments. After due deliberation and inspection *in loco* where necessary, the committee's recommendations are submitted to the Administrator for whatever action he may deem necessary. Schemes that have not been supported by the committee were the proposed developments at Sandy Bay and Capri East.

It soon became apparent that the efficiency of the advisory committee was being handicapped by three major factors, viz. the multiplicity of ownership, the shortage of funds and the fact that the committee is purely advisory and has no executive authority. Nevertheless, since its establishment in 1983, the committee has been reasonably successful in the conservation of the area, fostering a spirit of cooperation between the individual landowners, combating soil erosion and invasive aliens, improving the system of footpaths, curbing development and the over-utilisation of the nature area, and promoting conservation in general. The committee has recommended uniform legislation relating to the entire protected nature area as essential. For example, regulations relating to dogs in state forests differ from those on city council land, causing confusion among the public.

Apart from the perpetual threat of development, most of the problems with which the management committee

Litter at a picnic site.

Graffiti on a rock face.

is concerned arise from increasing pressures of human numbers and deplorable behaviour by the public. Whereas in the early days the mountain was used largely by serious mountaineers and members of the Mountain Club, who observed a strict code of conduct, it has now become a popular playground for more and more people of all ages, some of whom have little appreciation for nature. Existing paths are becoming eroded through overuse, new paths are trodden out and vegetation is trampled, resulting in soil erosion. Picnic spots are being degraded with litter, empty bottles and cans; trees and shrubs are mutilated; rocks are defaced with graffiti; and the silence of the mountain is polluted by transistor radios. The latest invader of the mountain is the so-called ATB (all-terrain bicycle). The use of bicycles in large numbers on the mountain will accelerate the erosion of steep paths and pose a threat to the safety of hikers. Trail or mountain bikes should therefore be restricted to existing roads and jeep tracks.

The fact must be recognised, however, that pressure on the mountains of the Cape Peninsula will continue to increase in the years to come, due to the alarming growth of the population of the Cape metropolitan area, the active promotion of tourism, the demands of developers, and the growing use of Table Mountain in particular for publicity purposes. It is estimated by the Human Sciences Research Council that there will be about four million people living in the metropolitan area of greater Cape Town by the turn of the century and their recreational needs will have to be met.

The use of Table Mountain for advertising purposes, publicity stunts or fund-raising, no matter how laudable the cause, should not be permitted, for at least two reasons. Table Mountain is part of our national heritage and as such should be regarded as sacrosanct. Secondly, the large number of spectators attracted by such events would result in extensive trampling of the vegetation, littering and general degradation of the environment. There is a limit to the number of people that can be accommodated in any nature area without the environment being degraded or the quality of the outdoor experience being reduced. As the CPPNE is accessible from all directions, it is not possible to control the number of people visiting it. Other methods will have to be employed to reduce the impact of users on these mountains.

Firstly, regional services councils and municipalities should make a major effort to provide large and attractive urban parks and recreational areas in the townships and suburbs. Many people who visit the mountains do so to escape from their dreary environment and to spend a day in the outdoors. Many visitors to Kirstenbosch, for example, have little interest in the flora, but go there simply to spend a few hours picnicking in pleasant surroundings.

To reduce trampling of the fynbos, an adequate, consolidated system of pathways must be provided and regulations promulgated to oblige visitors to keep to them. Notice-boards should be erected at all the major entry points to alert the public to the fact that they are entering a nature area and to inform them of the regulations applicable therein.

The Hey Commission strongly recommended that an adequate number of rangers be appointed to patrol the nature area, serving both as law enforcement and education officers, but due to a lack of funds this has never been implemented. While a great deal can be achieved by educational programmes, the mere fact that one may meet a uniformed ranger on the mountains would be a decided deterrent to would-be lawbreakers. To be effective, education must be supplemented by law enforcement, and the appointment of additional rangers is a matter of urgency.

Pressures for housing development both within the nature area and on the perimeter are likely to increase. Each application should be considered on its merits. There should be no large-scale development within the nature area, and any buildings permitted should blend in with the surroundings.

During 1993 three major initiatives were launched by the advisory committee to strengthen the conservation of the Cape Peninsula mountains. In order to reduce the problems hampering effective management arising from multiple ownership, the Administrator appointed a commission under the chairmanship of advocate Frank Kahn, to negotiate the rationalisation of the management of the area under fewer authorities. Agreement has been reached that the entire area will now be managed by the two major landowners, the Cape Town City Council and the Western Cape Regional Services Council. The exact areas have still to be determined, but it has been agreed that the City Council will be responsible for the Signal Hill and Table Mountain complex while the Regional Services Council will assume control from Constantia Nek to Cape Point.

As already mentioned, inadequate funding has always been a major factor in stultifying the effective management of the Peninsula mountains. After the disastrous veld fire on Table Mountain in December 1986, the then Minister of Environment Affairs, the late Mr John Wiley, convened a public meeting of representatives of the Department of Environment Affairs, local authorities, landowners, conservation organisations and the press, to discuss proposals for the improved management of the nature area. The meeting was attended by 65 delegates from 24 state and non-government organisations. Many proposals were made in the course of the discussions, and these were consolidated into a report for consideration by the management committee.

Most problems, it was found, arose from inadequate funding, the lack of law-enforcement personnel, and poor public behaviour. During 1972-1973 there were 497 fires on the mountain resulting from human agency. At the meeting the Minister announced the launch of a fund by the *Cape Times* to support essential management projects. The Southern African Nature Foundation accepted the responsibility of acting as trustee of this fund and of contributing to it on a rand for rand basis.

Over R55 000 was raised and allocated to the following works: the consolidation of footpaths; the control of alien vegetation; the purchase of equipment for use by hack groups; the design and erection of notice-boards at popular entry points to the nature area; and the preparation of educational posters. Recently the Nature Foundation has become more deeply involved in the conservation of Table Mountain and launched the Table Mountain Appeal with a target of R25 million, the interest from which will be used to fund management projects.

The third initiative of the advisory committee was to recommend that the University of Cape Town be commissioned to prepare a report on the future development and management of the mountain chain, the constituent nature reserves, monuments, forests and other land included in the Cape Peninsula Protected Natural Environment. This policy document is intended to provide the present or new structures of authority with guidelines on managing this mountain system more effectively with a view to conserving nature yet making the mountains accessible to the public as far into the future as possible. The document was presented to the present Management Committee in July 1994 and after due deliberation submitted to the Premier of the Western Cape Province together with the committee's recommendations. In brief the primary conclusions and recommendations of the consultants include:

- The entire Cape Peninsula requires special protection. Fundamental threats are the effects of increasing human population and unregulated urban expansion; alien biotic invasions; too frequent and widespread wild fires; air pollution; and divided control. Insufficient executive function and the plethora of controlling authorities having jurisdiction in the Cape Peninsula Protected Natural Environment further hinder effective management.
- The area at present called the Cape Peninsula Pro-

tected Natural Environment should be expanded to include the urban fringe and surrounding seas, and become known as the Cape Peninsula Heritage Area.

- A statutory authority, provisionally called the Cape Peninsula Heritage Area Authority, should be created by national legislation to act as the custodian of the Cape Peninsula, and to promote enhanced nature-orientated multipurpose use of the area.
- Until the Cape Peninsula Heritage Area Authority is legally constituted, it is recommended that the National Parks Board assume responsibility for the existing Cape Peninsula Protected Natural Environment and expanded area.
- Use of the Cape Peninsula should be managed so as to maximise societal benefits and to minimise environmental losses on a sustained basis.
- A properly structured campaign should be designed and instituted to enhance awareness among the public of the asset of the Cape Peninsula.
- Any changes to land, water and air in the Cape Peninsula should be regulated in such a way as to maintain and, where possible and appropriate, enhance the area's rich biological diversity.
- Human use of the Cape Peninsula should take place according to the management objectives of different functional areas, permitted uses varying from area to area. Five categories of functional areas are recommended with different intensities of prescribed use, ranging from low levels of human use to areas where intensive use would be desirable and acceptable (Environment Unit 1994).

While this report is an important milestone and a major step forward in the more effective management of the Peninsula Mountain Chain it must not be regarded as a magic formula to solve all the problems. The management of the CPPNE is, and will always be, a most difficult assignment, and especially so if the area is enlarged to include the urban fringe, as recommended. There is no simple solution to the manifold problems involved due to the multiplicity of ownership, including some 160 private properties comprising over 20 per cent of the total area, the impact of ever-increasing human numbers, and the poor behaviour of some members of the public. The successful management of this unique natural asset will depend largely on three factors: adequate funding by the state, local authorities, and private enterprise; more effective law enforcement; and the cooperation of the general public. Unless all members of the public using the nature area observe the basic rules of proper outdoor behaviour, no funding or management plans can ensure its preservation. The public must appreciate the fact that, individually and collectively, they are the stewards and custodians of this natural heritage. A single selfish or careless act can nullify many years of conservation work.

All users of the mountain should observe a strict code of behaviour. Such a code should be based on the principles of respect for the fauna and flora and the rights of others to enjoy unspoiled nature. The following draft code was prepared in consultation with the Southern African Nature Foundation, the Botanical Society and the Wildlife Society, and approved by the management committee. It is hoped it will be widely adopted by all interested organisations.

A mountain ethic

Access to nature is a privilege and implies a responsibility towards the conservation of the fauna, flora and scenic beauty, the heritage of future generations.

PLEASE RESPECT

THE MOUNTAINS

by staying on footpaths, taking your litter home, and not committing acts of vandalism, polluting waters or lighting fires.

THE VEGETATION

by not trampling it, collecting plants, flowers or seeds, or mutilating trees.

THE WILDLIFE

by not killing, collecting or disturbing in any way.

OTHER VISITORS

by not rolling rocks, throwing stones, playing radios or tape recorders, shouting or being ill-mannered.

Cherish the mountains and help to conserve nature so that it may also be enjoyed by future generations.

Since the days of the first free burghers at the Cape, South Africans have adopted the principle of the absolute rights of the landowner. The results of this policy are self-evident in the general appearance of our country today. Eroded farmlands, denuded mountain slopes, degraded pastures, vast dongas, drained wetlands, polluted waters, and overexploited natural resources. If we are to salvage viable representative remnants of our once rich heritage of natural resources of fauna, flora and incomparable scenic beauty, we need to adopt a new ethic – the ethic of "stewardship", not "ownership!" Mankind does not own the land, water or natural resources of

An eroded landscape.

the earth. We are merely granted the temporary use thereof. It is our solemn duty to pass these on unimpaired to future generations by observing a simple basic principle of nature conservation, viz.: Utilise natural resources but only within their natural regenerative capacity!

The Cape Peninsula Nature Area is unique, not only for its scenic beauty and remarkable flora and fauna but also because it lies, still relatively unspoilt, within a vast and growing metropolitan complex. It is a priceless natural resource, and it is hoped that this book will stimulate public awareness of its features and their importance, and help to gain the support of all sectors of the public for its conservation. Once destroyed, nature can never be restored!

The scenic beauty of mountain walks. (*Photo: W P U Jackson*)

References

Human history of Table Mountain

DR EDNA BRADLOW

1. R Raven-Hart, *Before Van Riebeeck. Callers at South Africa from 1488 to 1652*, (Cape Town, Struik, 1967), pp. 153-4, p. 28.
2. W Burchell, *Travels in the Interior of Southern Africa, Vol. 1*, (London, Longman, 1822), p. 8.
3. Raven-Hart, *Before Van Riebeeck*, p. 85.
4. Raven-Hart, *Before Van Riebeeck*, pp. 118, 141.
5. Raven-Hart, *Before Van Riebeeck*, facing p. 132.
6. F Valentyn, *Description of the Cape of Good Hope, Part 1*, P Serton (ed.), (Cape Town, Van Riebeeck Society, 2nd series, No. 2, 1971), p. 69.
7. M Murray, *Under Lion's Head*, (Cape Town, Balkema, 1964), p. 101.
8. R Elphick, *Kraal and Castle. Khoikhoi and the Founding of White South Africa*, (New Haven and London, Yale University Press, 1977), p. 91.
9. Raven-Hart, *Before Van Riebeeck*, pp. 15, 49, 119, 122; C P Thunberg, *Travels in Europe, Africa and Asia made between the years 1770 and 1779, Vol. 1*, (2nd edition, London, R and C Rivington, 1795), pp. 284-5.
10. J F Victorin, *Travels in the Cape. The years 1853-1855*, (Cape Town, Struik, 1968), p. 12.
11. R Semple, *Walks and Sketches at the Cape of Good Hope*, (London, C and R Baldwin, 1805), p. 14.
12. Thunberg, *Travels*, 1, p. 131.
13. Valentyn, *Description*, 1, p. 65.
14. R Ross, "The Occupations of Slaves in 18th Century Cape Town", (University of Cape Town, Cape Town Historical Conference, 1978).
15. A Sparrman, *A Voyage to the Cape of Good Hope, Vol. 1*, V S Forbes (ed.), (Cape Town, Van Riebeeck Society, 2nd series, No. 6, 1975), pp. 71-2.
16. L Guelke, "The White Settlers 1652-1780", in R Elphick and H Giliomee (eds), *The Shaping of South African Society, 1652-1820*, (Cape Town, Longman Penguin, 1979).
17. Valentyn, *Description*, 1, p. 199.
18. Thunberg, *Travels*, 1, p. 281.
19. Sparrman, *A Voyage*, 1, p. 66, n. 135.
20. A Trollope, *South Africa, Vol. 1*, (London, Dawsons, 1968), p. 83.
21. Burchell, *Travels*, 1, p. 29.
22. G21-1902, Cape of Good Hope. Report of a commission appointed to enquire into and report upon certain matters affecting The Cape Peninsula, (Cape Town, Government Printers, 1903).
23. J G Lockhart and C M Woodhouse, *Rhodes*, (London, Hodder and Stoughton, 1963), p. 203.
24. Trollope, *South Africa*, 1, p. 83.
25. "A Lady" (Mrs Louisa Ross), *Life at the Cape a hundred years ago*, (Cape Town, Struik, 1963), p. 9.
26. Thunberg, *Travels*, 1, pp. 221-2.
27. Raven-Hart, *Before Van Riebeeck*, pp. 19-20, 82, 118, 151.
28. Burchell, *Travels*, 1, p. 33.
29. C A Lückhoff, *Table Mountain. Our National Heritage after Three Hundred Years*, (Cape Town, Balkema, 1951), p. 14.
30. "A Lady", *Life at the Cape*, p. 27.
31. G21-1902, Report of a Commission, p. 12 ff.
32. W J Talbot, "Town and Country, Past and Present", in J A Mabbutt (ed.), *The Cape Peninsula*, (Cape Town, Maskew Miller, 1952), pp. 148-9.
33. E E Mossop, *Old Cape Highways*, (Cape Town, Maskew Miller, n.d. post-1927), p. 11.
34. W K Hancock, *Smuts, Vol. 2, The Fields of Force 1915-1950*, (Cambridge, University Press, 1968), p. 303.
35. Thunberg, *Travels*, 1, pp. 288-9.
36. Sparrman, *Voyage*, 1, p. 70.
37. Burchell, *Travels*, 1, p. 38.
38. R Ross, *Cape of Torments. Slavery and Resistance in South Africa*, (London, Routledge and Kegan Paul, 1983), pp. 19-20, 55 ff.
39. Semple, *Walks and Sketches*, p. 82; also Burchell, *Travels*, 1, p. 32.
40. *The Life and Adventures of Joshua Penny*, (Cape Town, S A Library, 1982), p. 35.
41. *Die Burger*, Bylae, 29.5.1976.
42. Raven-Hart, *Before Van Riebeeck*, p. 78.
43. B Warner, *Astronomers at the Royal Observatory Cape of Good Hope*, (Cape Town, Balkema for the University of Cape Town, 1979), p. 10.
44. Warner, *Astronomers*, p. 50.
45. Warner, *Astronomers*, p. 57.
46. Sparrman, *A Voyage*, 1, p. 53.
47. Thunberg, *Travels*, 1, p. 219 ff.
48. Thunberg, *Travels*, 1, pp. 264-269.
49. Thunberg, *Travels*, 1, pp. 291-2.
50. Burchell, *Travels*, 1, pp. 17-19, 34, 37.
51. Raven-Hart, *Before Van Riebeeck*, p. 8, quoting Barros.
52. Valentyn, *Description*, 1, p. 61.
53. Raven-Hart, *Before Van Riebeeck*, facing p. 132, p. 141.
54. B and N Warner, *Maclear and Herschel Letters and Diaries at the Cape of Good Hope 1834-1838*, (Cape Town, Balkema, 1984), p. 63.
55. J Murray, *In Mid-Victorian Cape Town*, (Cape Town, Balkema, 1953), p. 44; "A Lady", *Life at the Cape*, p. 42.
56. Raven-Hart, *Before Van Riebeeck*, p. 43, John Jourdain.
57. Valentyn, *Description*, 1, p. 61.
58. B and N Warner (eds), *The Journal of Lady Jane Franklin at the Cape of Good Hope*, (Cape Town, Friends of the S A Library, 1985), p. 58.

59. Semple, *Walks and Sketches*, p. 83.
60. Thunberg, *Travels*, I, p. 221.
61. Raven-Hart, *Before Van Riebeeck*, p. 118.
62. Murray, *Mid-Victorian Cape Town*, p. 41.
63. Semple, *Walks and Sketches*, p. 77 ff.
64. W H Crump, "Table Mountain", in A H Honikman (ed.), *Cape Town. City of Good Hope*, (Cape Town, Howard Timmins, 1966), p. 108.
65. Burchell, *Travels*, I, p. 33.
66. Burchell, *Travels*, I, p. 38.
67. "A Lady", *Life at the Cape*, p. 38
68. Hancock, *Smuts, Vol. 2*, p. 15.

Bibliography

Human use of the mountain
JOSE BURMAN

Alphine Club Journals: 1978-1987.
Barrow, J. 1801. *Travels into the interior of Southern Africa*. London: T Cadell & W Davies.
Bird, W. 1823. *State of the Cape of Good Hope in 1822*. London: John Murray.
Burman, J. 1966. *A peak to climb*. Cape Town: Human & Rousseau.
Burman, J. 1969. *Cape of good intent*. Cape Town: Human & Rousseau.
Burman, J. 1991. *The Table Mountain book*. Cape Town: Human & Rousseau.
Cape Argus, 29 August 1887.
Cape Argus, 2 December 1887.
Cape Times, September 1891.
Deeds Office Records.
Fairbridge Index.
Gubbins Library: Secret Memorial on Defence of the Cape by Sir James Craig.
Juta, R. 1927. *The Cape Peninsula*. London: The Bodley Head.
Laidler, P W. 1952. *A tavern of the ocean*. Cape Town: Maskew Miller.
Le Vaillant, F. 1790. *Travels into the interior parts of Africa* ... London: Robinson.
Lowe, C van Riet (ed). 1941. *The monuments of South Africa*. Pretoria: Government Printer.
MCSA Journals: 1893-1992.
Raven-Hart, R. 1967. *Before Van Riebeeck*. Cape Town: Balkema.
Raven-Hart, R. 1971. *Cape of Good Hope*. Cape Town: Balkema.
Robinson, Dr A M Lewin. 1973. *The letters of Lady Anne Barnard* ... Cape Town: Balkema.
SASA Bulletins: 1957-1960.
Table Mountain Aerial Cableway Company: Archives.
Theal, G M. *Cape Colonial Records 1795*.
Theal, G M. 1964. *History of South Africa*. Cape Town: Struik.
Theal, G M. 1964. *Records of south-eastern Africa*. Cape Town: Struik.
Van Riebeeck, Jan. 1958. *Journal of Jan van Riebeeck*, edited by H B Thom. Cape Town: Balkema.

Invertebrate fauna of the Cape Peninsula mountains
DR VINCENT WHITEHEAD

Claassens, A J M & Dickson, C G C. 1980. *The butterflies of the Table Mountain range*. Cape Town: Struik.
Coley, F. 1977. The fascinating caves of Table Mountain. *African Wildlife* 31 (no. 4): 20-22.
Dickson, C G C & Kroon, D M (ed). 1978. *Pennington's butterflies of Southern Africa*. Johannesburg: AD Donker.
Eastwood, E B. 1978. Notes on the scorpion fauna of the Cape. Part 3. Some observations on the distribution and biology of scorpions on Table Mountain. *Annals of the South African Museum* 74 (no. 10): 229-248.
Filmer, M R. 1991. *Southern African spiders: An identification guide*. Cape Town: Struik.
Lawrence, R F. 1984. *The centipedes and millipedes of Southern Africa: a guide*. Cape Town: Balkema.
Migdoll, I. 1987. *Field guide to the butterflies of Southern Africa*. Cape Town: Struik.
Moll, G. 1987. *Table Mountain, a natural wonder*. Cape Town: Western Cape Branch, Wildlife Society of Southern Africa.
Newlands, G. 1987. *Spiders and scorpions of Southern Africa*. Cape Town: Struik.
Newlands, G. 1986. *Struik pocket guides for Southern Africa: Spiders*. Cape Town: Struik.
Prins, A J & le Roux, V. 1986. *South African spiders and scorpions*. Cape Town: Anubis Press.
Skaife, S H. 1979. *African insect life*. 2nd edition. Cape Town: Struik.

Vertebrate fauna of the Cape Peninsula mountains
DR DOUGLAS HEY

Roberts, A. 1951. *The mammals of South Africa*. Johannesburg: CNA.
Frandsen, J. 1982. *Birds of the South Western Cape*. Sloane Park: Sable.
Hey, D (ed). 1987. *Cape of Good Hope Nature Reserve*. Cape Town: Regional Services Council of the Western Cape.
Mabbut, J A. 1952. *The Cape Peninsula*. Cape Town: Maskew Miller.
Rose, W. 1950. *The reptiles and amphibians of Southern Africa*. Cape Town: Maskew Miller.
FitzSimons, V F M. 1970. *A field guide to the snakes of Southern Africa*. London: Collins.
Wager, V A. 1965. *The frogs of South Africa*. Cape Town: Purnell.

A walker's paradise
JOSE BURMAN

Burman, J. 1979. *Latest walks in the Cape Peninsula*. Cape Town: Human & Rousseau.
Burman, J. 1984. *Where to walk from the cableway*. Cape Town: Human & Rousseau.

Conservation of the mountains of the Cape Peninsula
DR DOUGLAS HEY

Environment Unit. 1994. *Policy for multipurpose use of the Cape Peninsula*. Cape Town: University of Cape Town.

Index